OIL AND GAS ENVIRONMENTAL ECOLOGY

YURY I. PIKOVSKIY
NARIMAN M. ISMAILOV
MARINA F. DOROKHOVA

Translated (from Russian) "Prima Vista" firm, Russia

Editor of the English translation:
Dmitry E. Konyushkov

ACADEMUS
Publishing

Academus Publishing
2019

ACADEMUS
Publishing

Academus Publishing, Inc.

1999 S, Bascom Avenue, Suite 700 Campbell CA 95008
Website: www.academuspublishing.com
E-mail: info@academuspub.com

The right of Yury I. Pikovskiy, ScD in Geography,
Lomonosov Moscow State University, Department of Geography, Moscow, Russia;
Nariman M. Ismailov, ScD in Biology,
Professor, National Academy of Science of Azerbaijan, Institute of Microbiology,
Baku, Azerbaijan;
Marina F. Dorokhova, PhD in Biology,
Lomonosov Moscow State University, Department of Geography, Moscow, Russia.

Translation by "Prima Vista" firm, Russia.

Editor of the English translation:
Dmitry E. Konyushkov, PhD in Soil Sciences,
Russian Academy of Sciences,
V.V. Dokuchaev Soil Science Institute, Moscow, Russia.

Reviewers:
Alexander P. Khaustov, ScD in Geology,
Professor, RUDN University, Moscow, Russia;
Sergey V. Goryachkin, ScD in Geography,
Russian Academy of Sciences, Institute of Geography, Moscow, Russia.

ISBN 10: 1 4946 0014 5
ISBN 13: 978 1 4946 0014 3
DOI 10.31519/monography_1165

This book covers the fundamental problems of the interaction of hydrocarbons with the biosphere. It is based on long-term original studies by the authors and on information from modern scientific sources. Common features of carbonaceous substances — oil, natural gas, nature hard bitumen, and petroleum products — their chemical composition and toxicity are analyzed, and the main manifestations of the petroleum-driven anthropization of the environment are considered. The issues of stability of the natural systems in case of pollution by oil and petroleum products; the means for natural cleaning and remediation; and the methods for diagnostics, monitoring, and forecasting environmental changes caused by petroleum pollution are also discussed.

This book targets a wide range of readers and, particularly, students interested in the ecological role of oil and gas in the environment.

DEDICATION

MARIA A. GLAZOVSKAYA (1912–2016)

Maria Alfredovna Glazovaskaya is a distinguished Russian scientist and educationalist, geographer, pedologist, geochemist, professor, the founder of the Department of Landscape Geochemistry and Soil Geography at the Lomonosov Moscow State University. M.A. Glazovskaya made a significant contribution to the development of a new science — landscape geochemistry and exploration geochemistry; her seminal works covered a wide range of problems in the fields of soil formation, soil classification, geography of soils, and geochemistry of technogenesis. M.A. Glazovskaya created methodological basis of the oil and gas environmental ecology, which encompassed forecast landscape-geochemical zoning in terms of transformation of the natural environment by oil production and transportation with the determination of the environmental resilience to anthropization.

INTRODUCTION

Hydrocarbons are a global phenomenon on our planet. In the biosphere, hydrocarbons are ubiquitous, and appear in a variety of forms. Archaeologists found traces of oil and asphalt (bitumen) seeps on the Earth's surface that occurred many thousands of years before humans. Even back in antediluvian times, humans perceived numerous oil occurrences as part of the environment. Noah, before the Flood, used the pitch (tar) to seal to waterproof his giant ark, while the people of Babylon, who wanted to make a tower so high that it should reach the sky, used asphalt to cement the brickwork. They did not think whether the reserves of this material were sufficient to finish the job. Petroleum and gas have been escaping to the Earth's surface for thousands of years in many areas of the globe. However, no evidence exists that in the early days hydrocarbons were the biosphere antagonists. They were being dispersed in the atmosphere, consumed by microorganisms, oxidized and involved in photosynthesis as CO_2. The quantity of escaping hydrocarbons met the needs of the biosphere.

In today's world, oil and natural gas are the main fossil fuels. Since the 19th century, demand for these resources has been continuously growing. As petroleum is ubiquitous, it has been found in virtually all the regions, where oil exploration has been performed. Today, petroleum and gas are extracted from the Earth across all bioclimatic zones of the world on all inhabited continents, and in the surrounding seas. Earth holds enormous oil and natural gas resources. Although, they are distributed unevenly. Today's oil production rate is 30 billion barrels per year. Gas production rate is about 4 trillion cubic meters per year. Oil and, to some extent, gas serve as a feedstock for the petroleum products that are sold and used everywhere across the globe. Such expansion of hydrocarbons into the biosphere has gradually come into collision with the naturally established ecological balances. Oil, natural gas, and products of their processing now have a global negative impact on the environment, including the Earth's atmosphere, soil and vegetation cover, surface and groundwater on the continents, and the World Ocean.

Scientists from all around the world conduct a huge number of scientific researches aimed at studying the conflicts between petroleum production and the biosphere, and finding ways to protect it in the context of growing demand for hydrocarbon resources. Scientists and petroleum companies in many countries across the world are engaged in this activity. Results of their research are published in many languages.

Joint efforts of the scientists help create a separate field of knowledge about how petroleum and gas production can co-exist with the environment and the biosphere in general. It is called Oil and Gas Geoecology or Oil and Gas Environmental Ecology.

M.A. Glazovskaya greatly contributed to this field of knowledge via introducing general principles of landscape geochemistry (environmental geochemistry) into it [GLAZOVSKAYA, 1972, 1981, 1982, 1981, 1983, 1988, 2007].

The authors attempt to illustrate fundamental concepts of oil and gas geoecology, identify its most common problems and methods, and present preferred ways to protect and restore the environment affected by petroleum production. It is rather impossible to delve into the details of specific problems of oil and gas geoecology in the scope of a single book. Numerous studies addressing separate aspects of this science have been published. For details of these studies, see the extensive bibliography.

The book is a summary of the authors' long-term research, as well as their experience in teaching the fundamentals of oil and gas geoecology at Lomonosov Moscow State University and National Academy Science of Azerbaijan.

The first part of the book deals with common features of natural and anthropogenic carbonaceous substances chemical composition toxicity, specifically, oil, natural gas, nature hard bitumen, and petroleum products. The major focus is on the natural hydrocarbon flows in the biosphere. The Earth larger geochemical hydrocarbon cycle is considered. Hydrocarbon flows emerge and circulate not only in the biosphere but also in deep geospheres (the lithosphere, mantle, Earth's core), as well as in cosmic space. In the biosphere, these flows intersect. Natural hydrocarbons and carbon aceous substances in general are involved in the natural evolution of the biosphere without exposing it to any significant damage, if their simultaneous emissions remain below the critical threshold. A greater hazard for wildlife could come from some permanent geochemical associates that accompany hydrocarbons, such as high-salinity water, hydrogen sulfide, mercury, heavy metals, and radionuclides. Studying natural hydrocarbon flows in the biosphere helps better understand the role of the human-made carbonaceous substances that pollute the environment.

The second part of the book deals with the main manifestations of petroleum anthropization in the environment. The term petroleum anthropization describes geochemical processes related to the consequences of petroleum production environmental footprint (exploration, production, transportation, oil and gas processing, and use of petroleum products).

The third part of the book discusses issues of stability and regeneration of land and marine natural systems affected by petroleum anthropization. The biosphere has an enormous potential to withstand external impacts and natural purification. Mechanisms of natural purification of environmental components (soils, biocenoses, surface water, and groundwater) should be used to deal with the negative consequences of the environmental impact of the petroleum industry on ecosystems.

The fourth part of the monograph provides a brief description of the oil and gas geoecology methods that include diagnosis, monitoring, and forecasting environmental changes caused by petroleum anthropization. Particular attention is paid to the fluorescence analysis and bioindication of polluted environments, types, and origin of hydrocarbon geochemical fields in the soil cover and methods of forecasting environmental changes in a variety of physical and geographical conditions.

Most references to studies in the field of oil and gas environmental ecology concern Russian-language literature published in Russia, Azerbaijan, and other post-Soviet states. This literature is almost unknown to foreign readers. The authors' intention is to spark an interest in these studies that have made an unquestionable contribution to environmental ecology.

The authors hope that this work will be of interest to a broad audience of environmental protection specialists.

PART ONE
NATURAL AND ANTHROPOGENIC
CARBONACEOUS SUBSTANCES IN THE BIOSPHERE

CHAPTER 1
CHEMICAL COMPOSITION, PROPERTIES,
AND TOXICOLOGY OF CARBONACEOUS SUBSTANCES

1.1. THE CARBONACEOUS SUBSTANCES
AND THEIR COMPONENTS

Oil and natural gas are part of a large group of natural carbonaceous substances that are abundant in space, and all of Earth's spheres, including the biosphere.

Carbonaceous substances are generally non-crystalline molecular aggregates that vary in terms of their composition and origin, with the mass of carbon predominant over the mass of all other atoms. Apart from carbon, these aggregates contain hydrogen, nitrogen, sulfur, and oxygen on the macroscale, and almost all other chemical elements on the microscale. Natural carbonaceous substances include all classes of hydrocarbons and heteroatomic compounds, oil, natural gas, nature hard bitumen, all kinds of fossil coals, and oil shale. In the biosphere, natural carbonaceous substances exist in solid, liquid, and gaseous phases. Anthropogenic carbonaceous substances circulating in the biosphere appear in the same states. They are essentially products of oil, natural gas, coal, and oil shale industrial processing, as well as waste left after their economic use. Natural products released into the environment from technical facilities (wells, pipelines, tanks, machines) could be nominally included into anthropogenic carbonaceous substances.

Carbonaceous substances on the Earth mainly occur in two states, *concentrated state and dispersed state.* The bulk of carbonaceous substances in the atmosphere, soils, surface and groundwater, and rocks are in the dispersed state. Dispersed carbonaceous substances are ubiquitous in the biosphere and the lithosphere. In the lithosphere, concentrated accumulations of natural carbonaceous substances form fossil fuel deposits: oil, natural gas, nature hard bitumen, coals, and oil shale deposits.

All dispersed and concentrated carbonaceous substances and their components can be divided into *bituminous substances*, (soluble in organic solvents, mainly, in hydrocarbons and their chlorinated derivatives) and *carboids* (insoluble in organic solvents). The substance extracted from rocks, soils, and water bodies by means of a solvent is called a *bitumoid*. Natural bitumoids may include, apart from bituminous substances, lipid components of living or fossilized biota.

Natural and anthropogenic carbonaceous substances include four main groups of chemical components:

- natural hydrocarbon gases;
- liquid and solid hydrocarbons with admixed low-molecular-weight heteroatomic compounds;
- soluble high-molecular-weight substances (resins and asphaltenes);
- insoluble amorphous coaly substances with a polymer or polycondensation structure.

These components in different ratios produce the entire diversity of carbonaceous substances on the Earth.

Dispersed natural carbonaceous substance mainly consists of native amorphous carbon (carboids). In sedimentary rocks, these are mainly fossilized products of transformation of the residues of dead organisms buried in the course of sedimentation (*kerogen*).

Anthropogenic carbonaceous substances (petroleum products) consist of the same components as natural carbonaceous substances. However, the chemical composition of these components may differ from their natural counterparts.

Further, we will elaborate on the characteristics of carbonaceous substances being the primary objects of oil and gas environmental ecology: natural gases, oil, nature hard bitumen, and petroleum products. Coals and oil shales are studied by coal and shale environmental ecology.

1.2. HYDROCARBON NATURAL GASES

Natural gases permeate throughout the lithosphere, biosphere, and encircle the Earth in a continuous shell, the atmosphere. Occurrence forms and composition of natural gases on the Earth are diverse. Gas occurs in Earth's crust as concentrated accumulations in rocks filling the free space of rock pores and fractures. In high concentrations, gas may be dissolved in water or oil. In the dispersed state, natural gas is very common in soils; in sedimentary, igneous and metamorphic rocks; and in coal and saliferous beds. It is also released into the atmosphere during volcanic eruptions along with non-hydrocarbon gases.

Natural gases of oil and gas fields

On an industrial scale, natural gases are produced at gas, condensate, and gas-oil fields. Natural gases mainly consist of saturated hydrocarbons (> 95%). Their composition includes methane, ethane, propane, iso- and n-butane, pentane and its isomers, as well as small quantities of hexane, heptane, and heavier hydrocarbons, including aromatic and naphthenic compounds (benzene, toluene, xylene, cyclopentane, cyclohexane, etc.). Natural gases always contain some amount of non-hydrocarbon components. These are nitrogen, carbon oxides, water vapor, zero group elements (helium, neon, argon, etc.); hydrogen sulfide and other sulfur compounds (mercaptans). Hydrogen and vapors of mercury and fatty acids have also been detected in natural gases. Free gases either accumulate in the lithosphere or are discharged through high permeability areas onto the Earth's surface and escape into the atmosphere.

Solution gas is removed from oil after it is extracted to the surface using purpose-built plants, separators. This is associated gas. Generally, it is not commercially significant, and it used to be burned directly in the field using gas flares. This led to environmental pollution with products of its incomplete combustion. Today, the trend is to use this gas for utility purposes as much as possible.

Condensate. When natural gas occurs at great depths and under high pressures, light liquid hydrocarbons may be dissolved in it. In the course of gas extraction to the surface, lower pressure results in condensation of dissolved hydrocarbons into the liquid phase forming condensate.

Condensates differ from "regular" crude oil by having negligent or zero content of heavy components (resins, oils) in their composition, while the simplest petroleum components prevail. This is caused by the dissolving capacity of gases.

Dissolved natural gases

The lithosphere holds enormous hydrocarbon reserves in the form of gases dissolved in water.

These gases are divided into two major groups:
- gases dissolved in groundwater (water-dissolved gas);
- gases incorporated into the water structure that exist in the solid state at low temperatures (gas hydrate).

Water-dissolved gas. In terms of quantity, water-dissolved gases are by far the most common among biosphere gases. All gases dissolve in water, depending on their ubility factor. CO_2, H_2S, NH_3 easily dissolve in water. Natural gases, nitrogen, hydrogen, oxygen, as well as noble gases have low water solubility. Methane has the highest solubility among natural gases. Gas solubility increases along with the reservoir pressure. If the pressure decreases, some gas contained in the gas-saturated water can pass into the free phase and become the source of commercial production.

Pressure and temperature conditions, water-rock chemical reactions, radioactive decay, biochemical and other processes that are continuously taking place in all parts of the hydrosphere determine the diversity of water-dissolved gases.

Gas hydrates . Gas hydrates, especially, methane hydrates, stand out as a promising type of crude hydrocarbons. Gas hydrates are molecular compounds where one component (H_2O) forms a crystalline structure, whose cavities contain another component (C_nH_m). Gas-hydrate appearance resembles ice or wet snow. Gas-hydrate formula is $N \cdot nH_2O$, where N is the gas molecule, and n is the number of water molecules in hydrate crystals. This compound is stable at low temperatures and elevated pressure. For example, methane hydrate is stable at a temperature of $0\,°C$ and a pressure of around 25 bar and higher.

Gas hydrate is formed in deep permafrost areas, and deepwater areas of sea and ocean bottom sediments. Their primary areas of occurrence are confined to parts of the ocean bottom located at depths from 300 to 1,200 meters. Heating or pressure drop results in the compound decomposing into water and natural gas (methane). One cubic meter of methane hydrate produces 164 m^3 of natural gas under standard atmospheric pressure.

Gas-hydrate reserves on the planet are estimated to be at least 250 trillion m^3. In terms of calorific value, this is twice the value of all oil, coal, and gas reserves available on the planet combined [KVENVOLDEN, 1993]. However, a mass-production technology for gas-hydrate utilization is not yet available.

1.3. CRUDE OIL

Elemental and component composition of oil

Oil is a common name for naturally occurring multicomponent hydrophobic liquid solutions with a predominantly hydrocarbon composition which are contained in the free space of porous and fractured rock formations. Under conditions similar to the Earth's surface, liquid hydrocarbons act as oil solvents, while the substances dissolved in oil are gaseous and solid hydrocarbons, simple heteroatomic compounds, and high molecular weight substances — resin and asphaltenes. All hydrocarbon classes can be found in oil: saturated chain structures (alkanes and isoalkanes), mononuclear and polynuclear polymethylene, and aromatic structures. Heteroatomic compounds present in the oil composition include sulfur, oxygen and nitrogen compounds. Sulfur is a common chemical element in oil; its content varies from a few hundredths to 5–6% and higher.

Oil is an aggregate substance. All of its components occur in certain ratios. Oil's appearance, composition, and properties depend on these ratios. Today, oil is understood as a supramolecular assembly, consisting of many thousands of chemical structures connected by intermolecular bonds.

A complete list of individual compounds forming naturally occurring oil has not been established yet. More than one thousand individual hydrocarbons and simple heteroatomic compounds have been discovered in oil. Increasingly sophisticated analytical methods help discover new compounds annually.

Oil coming from any field across the globe is, on the one hand, similar in appearance and elementary composition and, on the other hand, occurs in a striking number of varieties. On the Earth's surface, petroleum is an oily liquid ranging from light yellow to dark brown in color with distinctly hydrophobic properties. A unique feature of oil existence in nature is the process of its continuous transformation by means of both internal potential energy and external factors. The elemental composition of tens of thousands of different individual oil samples across the globe varies in the range of just 3–4% (Table 1.1). Numerous trace elements contained in oil account for tenths or hundredths of a percent. Their list is virtually identical in any oil, although their quantity ratios may differ.

Table 1.1

Elemental composition of crude oil and hard bitumens
(Soboleva and Guseva, 2010)

Substances	Phases	C	H	N + S + O	Density, g/cm^3
Crude oil	Liquid	84–86	13–15	0.5–4	0.78–0.90
Maltha (heavy oil)	Viscous	80–87	11–12	3–7	0.95–1.05
Asphalt	Highly viscous, solid	78–87	9–11	5–10	1.05–1.10
Asphaltite	Solid	76–86	8–11	5–10	1.05–1.20
Kerite	Solid	80–90	4–9	5–10	1.10–1.30
Anthraxolite	Solid	90–99	4–10	0.5–5	1.30–2.0
Oxykerite	Solid	75–80	7–8	10–20	1.15–1.25
Huminkerite	Solid	65–75	5–8	20–30	1.25–1.50
Ozokerite	Solid	85–86	14–15	0.05–4	0.91–0.97

Table 1.2

Component composition (%) of oil and hard bitumens
(SOBOLEVA AND GUSEVA, 2010)

Substances	Naphtha	Resin	Asphaltene	Carboid
Crude oil	65–100	0–30	0–5	0
Maltha (heavy oil)	45–65	30–40	5–15	0
Asphalt	25–45	30–50	15–40	0
Asphaltite	5–25	5–50	30–90	0–10
Kerite	1–20	5–20	1–50	10–95
Anthraxolite	–	–	–	95–100
Oxykerite	1–5	1–5	60–80	10–40
Huminkerite	0–2	0–0,2	0–60	40–60
Ozokerite	95–100	race	trace	0

Nevertheless, oil is extremely diverse in terms of various hydrocarbon classes and the high molecular weight heteroatomic compounds ratio. It may differ not only between the fields, but also between different areas of the same field. Oil varies substantially in composition and the solvent / dissolved components ratio, its physical and chemical properties, and toxicity. Oil composition, properties, structure, and state are determined by the conditions of its formation (reservoir temperature and pressure, and surrounding rock composition).

Petroleum component composition is the weight percentage ratio of light and heavy hydrocarbons, resin, and asphaltenes. These components allow to differentiate between light crude and naphtha oil consisting mostly of hydrocarbon components and light resins, and resinous and resinous-asphaltenic oil where a higher concentration of resin and asphaltenes is observed (Table 1.2).

Light crude oil contains virtually no high-molecular-weight components. Its density is normally below 0.8 g/cm^3 (corresponding to very light crude), and its hydrocarbon composition mainly includes low-molecular-weight alkanes and cycloalkanes. This group includes all gas condensates as well.

Naphtha oil contains all classes of hydrocarbons, including polycyclic structures, and a small amount of low-molecular-weight resin. Asphaltenes are completely absent in this group. In terms of density, it is light and, partially, medium oil.

Resinous oil contains as much as 20% of resin and up to 5% of asphaltenes. Higher resin content results in higher molecular mass. The alkane ratio in the hydrocarbon composition reduces, while the significance of cyclic structures, especially, aromatic hydrocarbons, grows. In terms of density, this group includes medium and heavy oil.

Resinous asphaltenic (highly resinous) oil is referred to as heavy oil. It includes up to 50% of high-molecular-weight resins and asphaltenes, including up to 10% of pure asphaltenes. It is viscous slow-moving oil. Its hydrocarbon composition is dominated by cyclic and polycyclic structures.

Resinous substances are highly sensitive to free oxygen and can easily attach it. In the open air, resinous oil quickly thickens and loses its mobility. Probably, atmospheric oxygen plays an essential role in formation of new resins thanks to aromatic and hybrid structures. Access of microorganisms to these substances is restricted, and their metabolism process is very slow, sometimes it takes decades.

Fractional composition of oil

Light fractions, according to their boiling temperatures, are divided into lightweight (from boiling point up to 200 °C) and middleweight (200–350 °C). Lightweight fractions are referred to as gasoline fractions, whereas middleweight fractions are called kerosene-gas-oil fractions.

Oil components with a boiling temperature above 350 °C are heavy, or dark, fractions — residual fuel oil. Thermal distillation of residual fuel oil up to 550 °C requires vacuum. The residue boiling above this temperature is referred to as tar oil. Production process for a variety of petroleum products is based on thermal fractional oil distillation.

Oil hydrocarbon-type content

Depending on the hydrocarbon classes ratio in individual fractions and in oil in general, the following main classes of oil are distinguished: methane (paraffin), naphthenic, aromatic oil, as well as intermediate classes — naphthenic-methane, methane-naphthenic, aromatic-naphthenic, and naphthenic-aromatic.

To differentiate between these classes, A.F. Dobryansky [1961] suggested the following threshold values for content of various hydrocarbon classes (Table 1.3).

Table 1.3

Hydrocarbon classes of oil (DOBRYANSKY, 1961)

Index	Hydrocarbon classes	Alkanes, %	Naphthnes, %	Aromatic, %	Density, deg. API
1a	Aromatic	0–5	45–55	50–55	21–17
1b	Naphthenic-Aromatic	5–10	50–60	35–45	24–21
2a	Aromatic-Naphthenic	5–15	50–60	30–40	27–24
2b	Naphthenic	10–20	50–60	20–30	31–24
3a	Paraffin-Naphthenic	20–30	50–60	15–25	35–31
3b	Naphthenic-paraffin	30–40	45–50	10–15	39–35
4	Paraffin	40–55	35–45	5–10	45–39

Polycyclic aromatic hydrocarbons in oil

Polycyclic aromatic hydrocarbons (PAHs) may be regarded as important ingredients of intricate carbonaceous substances, and as geochemical markers of natural and anthropogenic substances fluxes in environment. The PAH structure consists of two or more condensed benzene rings (Fig. 1.1).

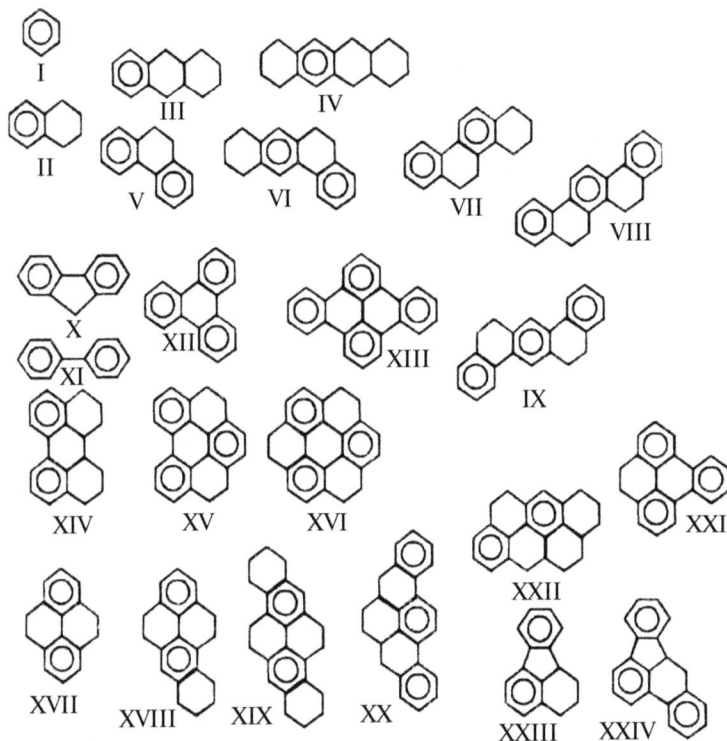

Fig. 1.1. Examples of structure of polycyclic aromatic hydrocarbons:
I — Benzene; II — Naphthalene; III — Anthracene; IV — Tetracene;
V — Phenanthrene; VI — Tetraphene; VII — Chrysene; VIII — Picene;
IX — Dibenzo(a,h)anthracene; X — Fluorene; XI — Biphenyl; XII — Triphenylene;
XIII — Dibenzo(a,h)pyrene; XIV — Perylene; XV — Benzo(ghi)perylene;
XVI — Coronene; XVII — Pyrene; XVIII — Benzo(a)pyrene;
XIX — Dibenzo(b,def)chrysene; XX — Benzo(rst)pentaphene;
XXI — Benzo(e)pyrene; XXII — Anthanthrene;
XXIII — Fluoranthene; XXIV — Benzo(b)fluoranthene.

The PAHs content in oil varies between 1% and 5% depending on the area. PAHs that occur in oil contain an alkyl chain instead of hydrogen atom in one or several radicals. For this reason, these molecules can be viewed as *substituted* homologues of the corresponding *unsubstituted* hydrocarbons. More than half of them (55%) are homologues of naphthalene with two aromatic rings; 27% are homologues of phenanthrene and anthracene with three aromatic rings; and 15% are homologues of pyrene, chrysene, and benzofluorene with four rings. The higher molecular weight PAHs ratio amounts to about 3% of their total quantity. Unsubstituted aromatic hydrocarbons rarely occur in crude oil and their quantity is insignificant. Information on benzo[a]pyrene content in oil is often ambiguous. Benzo[a]pyrene is rarely found in crude oil that had not been exposed to a strong thermal impact. However, its quantity rises sharply in oil refinery products. Some publications indicate that benzo[a]pyrene amount per 1 kg of oil may reach into the range of hundreds and thousands of µg. Nevertheless, normally the alkyl-substituted homologues of benzo[a]pyrene are present, and their carcinogenic activity is substantially reduced. Polycyclic arenes in oil are dominated by the compounds containing no more than three benzene rings in the molecule.

Trace chemical elements in oil composition

Besides carbon, hydrogen, nitrogen, sulfur, and oxygen, oil composition includes more than 60 chemical elements in trace concentrations (V, Ni, Co, Mn, Mo, Fe, Al, etc.). All known oils essentially contain the same set of trace elements, although their relative concentrations vary significantly depending on the age of rock surrounding the oil, regional tectonic conditions, and the composition of pure oil. Trace elements are mainly linked to the high molecular weight components of oil: resins and asphaltenes. Therefore, the more resinous oil, the more trace elements in its composition. Particularly large quantities of trace elements occur in nature hard bitumen, in which the high molecular weight part dominates over the hydrocarbon part. Nature hard bitumen is therefore used as an ore for extracting certain metals, for example, vanadium. Some trace elements are linked to oil itself and its formation conditions (V, Ni, Cu, Zn, Pb, Co, etc.).

Other trace elements could be transferred from oil-surrounding rocks and reservoir waters (Fe, Ca, Mg, Na).

The most widespread trace elements in oil are vanadium and nickel. Nitrogen-containing porphyrin complexes contain from 4 to 20% of vanadium and nickel present in oil. The remainder is present in other complex compounds. Vanadium predominates in sour crude oil and asphaltenes, while nickel is predominant in sweet crude oil and resins.

1.4. NATURE HARD BITUMENS (NAFTIDES)

Nature hard bitumens are multicomponent carbonaceous substances that closely resemble high molecular-weight oil components in their origin and chemical composition. Some nature hard bitumens were formed in the process of oil transformation under the influence of external factors (exposure to oxygen or microorganisms). Others were created in rock fractures as a result of precipitation from high-temperature gas solutions. Nature hard bitumens always occur as derived fossils, i.e., they fill up the cavities and fractures in the already formed rocks (see Tables 1.1 and 1.2).

Malthas (heavy oils) and asphalts are highly viscous, hard, and highly resinous carbonaceous substances. Malthas occur at depths of up to 3,000 meters; their largest reserves are found at depths between 800 and 1,600 meters. Asphalts are solid substances that have completely lost their mobility. Compared to oil, their elementary composition features higher content of oxygen, sulfur, and nitrogen. Accumulations of maltha and asphalt on the Earth can be enormous. Their deposits could stretch for thousands of kilometers, and their world reserves are several times the forecast global petroleum reserves. Several huge asphalt lakes are known on the Earth. The bottom part of those lakes consists of maltha, while their surface consists of asphalt (Sakhalin Island, Trinidad Island). More often, malthas and asphalts saturate porous rocks over large areas. The largest known asphalt deposits hold the geological reserves of tens of billions of tons (Lake Athabasca and Peace River in Western Canada). A thick mixture of asphalt and maltha cements the mass of tight sandstones near the surface of those deposits.

Usually, maltha and asphalt formation is related to oxidation, biodegradation, and light fractions loss that occur in resinous oil at shallow depths. However, sometimes, heavy oils occur at depths of 3 km and more. Several examples are known when large amounts of maltha and asphalt were forced from deep strata to the Earth's surface through fractures in Earth's crust ("asphalt volcanism").

Asphaltites are carbonaceous substances which composition by volume is dominated by asphaltenes (up to 90%) soluble in low-polar organic solvents. Asphaltenes typically occur as vein deposits. In terms of elementary composition, asphaltites are similar to asphalts. Asphaltite varieties include *grahamite and gilsonite.*

Kerites are highly carbonaceous substances that are only partially soluble in organic solvents. They occur as veins of varying, sometimes large, dimensions, or granular inclusions in rock. Kerite appearance is similar

17

to bituminous coal, and they used to be often described by the term "vein coal". Kerite varieties are albertite and impsonite. In albertites, bituminous substances dominate over carboids, while in impsonites, carboids dominate over the bituminous substance. Asphaltites and kerites are valuable as a raw material to produce rare metals and radioactive elements.

In terms of their combined properties, *anthraxolites* are similar to anthracite varieties of coal. They are insoluble in organic solvents. Carbon makes up 92–97% of the elementary composition. The density rises to 1.30–1.70 g/cm^3.

Ozokerites are wax-like substances that mainly consist of hard paraffins — saturated chain hydrocarbons. Formation of their deposits is linked to loss of gas and liquid components from gas-saturated paraffin-base oil near the surface. Normally, solid hydrocarbons are able of holding only a small amount of petroleum components on their surface.

Algarites and mumiyo, as well as oxykerites and huminkerites, could be also nominally included into the nature hard bitumen category. Algarites and mumiyo are water-soluble hydrocarbon-protein products of paraffin-base oil microbiological processing. *Oxykerites and huminkerites* are products of oil and bitumen oxidation and humification on the Earth's surface. Huminkerites are fully humified varieties. They are dissolved in alkalis.

1.5. PETROLEUM PRODUCTS

Petroleum products are carbonaceous substances of anthropogenic origin. They are mainly produced at large-scale oil-refining facilities. The industry produces hundreds of petroleum product types and thousands of brands that vary in composition, properties, and toxicity. The substance that are initially present in oil are extracted during its primary processing. Primary processing products are obtained by straight fractional distillation of oil: hydrocarbon solutions are separated into fractions with different boiling temperatures (Table 1.4).

In the course of secondary oil processing, primary distillation products are exposed to thermal and catalytic processes in order to improve feedstock quality and produce additional quantities of light petroleum products, as well as improve their consumer performance. Secondary processes are used to obtain petroleum products from heavy residue include catalytic reforming, catalytic cracking, visbreaking, thermal cracking, hydrocracking, coking of petroleum residue, pyrolysis, and others. During these processes, reactor temperatures can reach 500–600 °C and more. In addition to substances contained in naturally occurring oil, new toxic compounds, which are rarely found in nature, are

Table 1.4
Oil primary processing products

Boiling temperature range, °C	Resulting product
Atmospheric distillation	
140	Gasoline fraction
140–180	Ligroin fraction (heavy naphtha)
180–240	Kerosene fraction
240–350	Diesel fraction (light gasoil, straw dis-tillate)
Above 350	Residual fuel oil
Vacuum distillation	
350–420	Lightweight hydrocarbon oils (insulating oil)
420–500	Vacuum gas oil Medium- and heavy-weight hydrocarbon oil
Above 500	Tar oil (vacuum residue)

introduced. These include, in the first place, reactive unsaturated hydrocarbons, polycyclic aromatic hydrocarbons, and their derivatives with a carcinogenic and mutagenic effect on animal organisms and humans, and a variety of other non-hydrocarbon compounds.

Petroleum products can be divided into the following main groups, depending on their composition, properties, and application areas:

I — gaseous and liquid fuels;

II — grease;

III — paraffin, ceresin;

IV — bitumen;

V — carbon black (soot);

VI — petroleum coke, pitch;

VII — other special-purpose petroleum products (solvents and others).

Gaseous hydrocarbon products for domestic use are made using C_1–C_4 hydrocarbons obtained by primary distillation, catalytic reforming, or catalytic cracking. Gases are produced in grades with different propane and butane content. These products are limited by mercaptan sulfur and hydrogen sulfide content by weight (0.015% max.).

Motor fuel is the most widespread petroleum product, and it is more often responsible for pollution of environmental components as compared to other petroleum products. Motor and aviation gasolines are produced using oil light fractions with boiling temperatures up to 217 °C. Gasoline composition includes regular and isomeric structure alkanes. They are obtained by straight distillation of feedstock oil, whereby hydrocarbons are introduced from the catalytic cracking process designed to split the heavy molecules into smaller molecules. As a result, gasoline contains large quantities of chemically active unsaturated compounds. They are detrimental for fuel stability; hence, their amount is limited to 25%. In order to raise the gasoline octane number, isoalkanes, obtained by alkylation and isomerization process, as well as aromatic hydrocarbons, are added. Due to high toxicity of aromatic hydrocarbons, their content in gasoline is being continuously reduced. In Russia, this limit is down to 5%, while in foreign-made gasolines, it reaches 1% (DANILOV, 2003). Gasoline sulfur content is limited to 0.1%. Jet fuel (aviation kerosene) is primarily obtained by straight oil dislation (98% of fuel components boil off below 250–280 °C). Fuel grades differ by fractional composition, as well as total and mercaptan sulfur content (total sulfur 0.1–0.25%, mercaptan 0–0.005%). Freezing point for aviation kerosene must not be higher than −60 °C.

Diesel fuel. Light gasoil distills at temperatures of 200–360 °C, heavy gasoil — at temperatures of 360–500 °C. Composition: alkanes and cyclanes — 60–90%, arenes — 10–40%. Diesel fuel is designed to be used in compression ignition engines. It is produced from fractions of straight distillation and secondary processing of oil. Several diesel fuel grades are produced: fuel for high-speed diesel and gas turbine engines installed in ground vehicles and vessels (grades: S — summer fuel, W — winter fuel, A — arctic fuel). Fuel types differ in freezing temperatures, viscosity, and sulfur content. Diesel fuel environmental safety specifications are sulfur content — 0.05% (USA), 0.2–0.5% (Russia); arene content — 10% maximum. Additives, i.e. depressors based on ethylene-vinylacetate copolymer are introduced into diesel fuel to improve its quality, including to reduce its freezing temperature. Boiler fuel (residual fuel oil) is prepared from residues of various oil refining processes. Boiler fuel composition includes high molecular weight hydrocarbons, resinous-asphaltenic substances, and carboids. In terms of production method, fuel oil can be heavy oil residues of straight distillation or heavy high-viscosity cracking process residue. In the latter case, fuel can contain a large amount of polycyclic aromatic hydrocarbons. Boiler fuel is characterized by high sulfur content present in the form of organic sulfur compounds (up to 2.0–3.5%).

Grease includes high-boiling liquid fractions of petroleum products varying in viscosity and purification degree. They are primarily intended to reduce friction and wear of mechanical parts, transfer heat from the motor and heated parts, and protect parts from corrosion. Lubricating oil composition is represented by a mixture of high molecular weight methane, and naphthenic and aromatic hydrocarbons with a small quantity of resinous-asphaltenic substances. Lubricating oil composition for different applications is ensured by mixing base oils with a variety of additives.

Mineral base oils are obtained from tar oil or by vacuum distillation of residual fuel oil. Resulting oil fractions can be used for oil production only after they have been cleaned of foreign components and admixtures by means of solvents and sorbents. Polycyclic aromatic hydrocarbons, resinous-asphaltenic substances, petroleum acids, heteroatomic compounds that contain nitrogen, sulfur and oxygen, as well as certain metals, should be removed. Naphthenic hydrocarbons and isoalkanes have the most valuable properties for lubricating oil production.

Paraffin and ceresin group includes liquid paraffins, petroleum paraffin wax, food-grade paraffin wax, and ceresin. Liquid paraffins serve as raw material for producing protein-vitamin concentrates, synthetic fatty acids, and surfactants. Paraffin wax is separated from lube distillation fractions and is used for technical applications, food industry, and healthcare. Food industry paraffin is characterized by the fact that it is completely free of benzo[a]pyrene, acids, alkali, sulfates, chlorides, water, and mechanical impurities. Paraffins are not toxic to organisms. No conspicuous environmental issues related to industrial paraffin pollution have been recorded.

Anthropogenic bitumen consists of carbonaceous substances solid and semi-solid components: resin, asphaltenes, and carboids. Residual oil refining products are the primary feedstock for anthropogenic bitumen production: tar oil, or asphaltenes stripped during petroleum products deasphaltizing, or base oil purification. Two methods of bitumen production exist: (1) vacuum distillation of heavy highly-resinous oil and (2) oxidation of petroleum residues with oxygen or air. Bitumens are used in road construction, roofing materials, hydraulic insulation, electric insulation production, and asphalt varnish environmental importance is connected to their wide application in road construction.

Carbon black (soot) is a black dispersed powder. It is produced at special-purpose plants from pyrolysis resin, catalytic and thermal cracking gasoil, and natural gas at temperatures between 1,200 and 1,700 °C. It consists mainly of carbon (95–99.5%). This product contains small amounts of hydrogen (0.2–0.9%), sulfur (0.01–1.2%), oxygen (0.1–5%), and ash — up to 0.3%. Oil refining fractions with high arene and coke content, as well as natural gases, are the feedstock for carbon black production. Soot contains a large amount of various polycyclic aromatic hydrocarbons, of which many have a strong carcinogenic effect. Carbon black production involves formation of large amounts of carcinogenic polycyclic aromatic hydrocarbons that are released to the atmosphere with process emissions, and then penetrate into soils and water bodies. Benzo[a]pyrene concentration in soils surrounding carbon black production facilities exceeds the maximum permissible limit by a few dozen times. Up to 7 million tons of carbon black for a variety of applications is produced worldwide each year.

Petroleum coke and pitch are solid and semi-solid high molecular weight multicomponent carbonaceous products that are obtained by coking of petroleum residual fractions and thermal condensation of resinous-asphaltenic substances and condensed aromatic hydrocarbons. Petroleum coke and pitch almost do not decompose in the environment. They contain many carcinogenic polycyclic aromatic hydrocarbons, which makes them even more dangerous for the environment.

1.6. TOXICITY OF OIL, NATURAL GAS, AND PETROLEUM PRODUCTS

The impact of oil on living organisms

Chemical toxicity of oil for biological species is not always evident. It is known that small amounts of oil in certain cases produce a stimulating effect on plant growth. Oil is a nutritious substrate for a variety of groups of microorganisms. It degrades much more easily than many other toxic substances and injects with fresh additions of organic compounds into the soil.

Available data reveal that naphthenic oil produces a stimulating effect on living organisms. As an example, one could mention therapeutic oil from the Naftalan field in Azerbaijan. It exhibits a high biological activity and is used for treatment purposes. Naftalan oil is used for treatment of cardiovascular, musculoskeletal and nervous system, gastro-intestinal tract, skin, and many other diseases.

Polycyclic naphthenic structures are the ones responsible for the biological activity of oil. Oil stripped of aromatic and resinous compounds, almost colorless, and containing only naphthenic hydrocarbons, produces the most powerful balneotherapeutic effect. At the same time, some toxic properties have been also noted in Naftalan oil. They occur when light fractions containing methane hydrocarbons from deeper horizons are carried into the oil [GULIYEVA, 1981]. Naftusya mineral water at the Truskavets Resort, Ukraine, is known for its therapeutic properties. In terms of its composition, Naftusya is a calcium bicarbonate magnesium water, but it also contains the dissolved petroleum component and several biologically active substances present in oil.

Paraffin wax separated from oil and purified has been successfully employed in healthcare. However, paraffin wax in the open air is slow to degrade and oxidize. That is why if paraffin oil ends up in the soil, plants cease to receive moisture and nutrients, resulting in destruction of the entire ecosystem in the polluted area.

Oil toxicity is revealed when it penetrates into ecosystems. Mass mortality incidents that involved various plants caused by soil pollution by oil are well known. If oil is released in water, it becomes unusable for life and households. Living organisms that inhabit polluted water or that were grown on polluted soils lose their edible qualities. When oil and petroleum products penetrate soils and waters, they disrupt the geochemical balance established in the ecosystems. These disruptions are caused by several mechanisms: changes of the environment physical conditions, and its air and water regime; introduction of the toxic substances that inhibit the activity of certain bio-community components; changes of migration capabilities of individual chemical elements in the soil; soil salinization by associated mineralized reservoir water, formation of bituminous salt lakes. Plant growing on soils polluted by polycyclic aromatic hydrocarbons, substances with pronounced carcinogenic and mutagenic properties, is harmful for animals and is especially hazardous to human health.

This diversity of oil impacts on ecosystems depends on many factors. These are oil amount and composition, its degradation and dispersal rate, as well as composition of its permanent associates: mineralized reservoir water (up to the state of brine), toxic gases and volatile compounds (hydrogen sulfide, mercury, etc.), heavy metals, radionuclides, and PAHs. The effect of these factors ultimately determines stability of ecosystems in case of pollution by oil and petroleum products.

Whether an ecosystem adapts to new conditions and starts regenerating its functions or degrades depends on the above-mentioned factors. In general, one should pay attention to the diversity of combinations that include oil types and soil and bioclimatic conditions, in which it interacts with living organisms.

Toxic components in oil

Methane hydrocarbons of light oil fraction, when present in soils, water, or air environment, produce a narcotic and toxic effect on living organisms. Normal alkanes with a short carbon chain contained mostly in light oil fractions act especially fast. These hydrocarbons are more soluble in water and easily penetrate through membranes into living cells.

The nature of oil effect on living organisms is largely determined by its liquid component vapors. Oil with low content of aromatic hydrocarbons has an effect essentially similar to that of a methane and naphthenic hydrocarbons mixture. Its vapors cause unconsciousness and convulsions. High aromatic hydrocarbon content causes a risk of typical

23

chronic intoxications with pronounced changes in blood making organs and blood composition.

Aromatic hydrocarbons per se are a highly toxic component of oil and petroleum products. Raising their concentration in water and soil leads to plants inhibition or death. Oil and petroleum products enriched with aromatics are more toxic than paraffin oil. The PAH content in petroleum products can be substantially higher than in oil because of new formations created during thermal catalytic oil refining processes.

Contact with crude oil results in typical skin diseases. Sometimes even a single contact with sulfurous oil or its highly concentrated vapor can cause diseases. Intoxications and skin damage can occur in oil fields, oil refineries, when oil is used for ore flotation, as fuel, grease, dust containing agent, etc.

Sulfur compounds presence in oil significantly affects its quality and toxicity. Besides, they are highly corrosive. This is especially true for hydrogen sulfide and mercaptan. Higher oil sulfur content increases the risk of hydrogen sulfide contamination for polluted water bodies and overwatered soils (bog and grassland soils). Lighter fractions of sulfurous oil (up to 200 °C) often contain mercaptans that produce a strong, lasting unpleasant smell.

Toxic properties of petroleum products

Oil, its vapor, gases, and refinery products (gasolines, solvents, lubricating oils, paraffins, bitumens, petroleum coke, etc.) are highly toxic. The higher the content of sulfur-containing compounds in oil, the more toxic effect it produces.

The overwhelming majority of the substances used in oil refining and petrochemical industry are fire and explosion hazardous, toxic, and carcinogenic. Inside a body, petroleum products have a general toxic effect, with neurotoxic action especially pronounced. Inhaling hydrocarbon solvents creates the greatest hazard to health. Hydrocarbons inhibit reduction-oxidation processes and disrupt tissue respiration. They suppress cardiac activity, decrease anti-toxic liver function, secretory function of digestive organs, adrenal cortex, thyroid gland, and other functions.

Liquid fuel toxicity depends on the presence of light fractions, aromatics, and sulfur. Engine exhaust contains many polycyclic aromatic hydrocarbons, in particular, carcinogenic benzo[a]pyrene. Chemical additives used to increase fuel performance may contain toxic substances. Typically, soil pollution by various fuels occurs at oil terminals (filling operations, leakage due to tanks corrosion), on product pipelines, and at gas stations. Pollution area is not limited by the facility boundaries but extends far beyond.

Hydrocarbon oil carcinogenicity depends on the contents of polycyclic aromatic hydrocarbons. Benzo[a]pyrene has been found to be absent in most industrially produced hydrocarbon oils.

Toxicity of other oil refining products (plastic lubricants, fuel, and oil additives, etc.) has been studied in less detail. Their toxicity and nature of biological effect depend on the composition of individual chemical ingredients. Waste hydrocarbon oils are extremely detrimental to the environment. Many carcinogenic substances, including the most common one, benzo[a]pyrene, will form in this oil since it was exposed to temperatures up to hundreds of degrees inside mechanisms. Waste oils are a favorable growing medium for microorganisms, including pathogenic ones.

Hydrocarbon oils degrade in soils and sediments much slower that liquid fuel. Besides, some fuels undergo an oxidation and polycondensation process in the soil, which increases their viscosity, making them similar to oils in terms of diagnostic properties. Benzene has an irritating effect on upper respiratory tract mucous membrane. When sharply inhaled, it quickly accumulates in the central nervous system cells that are rich in lipoids, and produces a narcotic action. Prolonged exposure to relatively low concentrations inhibits the function of bone marrow.

Chronic intoxication with toluene is connected with its narcotic effect: it results in headache, dizziness, general weakness and asthenia, memory problems and loss of coordination, and nausea. Phenol (carbolic acid, hydroxybenzene), one of the most common environment pollutants, is a colorless crystalline substance with a distinctive odor. It is a strong poison that affects the nervous system and has a local burning action on the skin. Phenol intoxication occurs when the substance enters the gastrointestinal tract or penetrates through the skin.

Bitumen, solid and semi-solid oil distillation residue, is an extremely complex mixture that mainly consists of high boiling hydrocarbons. Some bitumens are carcinogenically active in case of skin contact. Vapor produced by asphalt and asphalt dust can lead to respiratory tract cancer.

1.7. TOXIC GEOCHEMICAL ASSOCIATES OF OIL AND NATURAL GAS

Toxic trace chemical elements

Certain metals present in resins and asphaltenes composition or absorbed by them can have a toxic effect on living organisms. In the process of oil field development, gases, waters of various composition, rock particles, their suspensions, slurries, bitumen, and oil components are lifted to the daylight surface. Harmful elements transferred during oil field development include Ba, V, S, Cd, Co, As, Ni, Hg, Pb, Sr, Zn, as well as radionuclides. In terms of environmental safety, oil trace elements can be divided into two groups: non-toxic and toxic. Non-toxic and low-toxic trace elements, which make up the largest portion of petroleum ash, are Si, Fe, Al, Mn, Ca, Mg, P. V and Ni are the most common toxic metals that are concentrated in resins and asphaltenes. Both metals are present in the composition of porphyrin complexes. Vanadium content can reach 40% of ash (0.04% of oil), nickel content — 16% of ash (0.01% of oil). High concentrations of nickel and, especially, vanadium compounds act as poisons. They suppress enzymatic activity and affectrespiratory organs, blood circulation organs, nervous system, and skin of humans and animals [YAKUTSENI, 2005].

In emissions of heat and power plants and oil refineries, trace elements bond with aerosols to reach high concentrations.

The quantity of trace elements with the highest concentrations in oil (vanadium, nickel, iron, copper, zinc, lead) is such that petroleum or fuel oil processing or combustion results in discharging significant quantities of toxic substances that contain these chemical elements into the environment. For example, modern cogeneration power plants running on sulfurous residual fuel oil daily discharge up to 1,000 kg of V_2O_5 with their smoke. Ash in these CPPs is much richer in vanadium than many industrial ores. In 1987 alone, more than 2,000 tons and more than 900 tons of vanadium were produced from heavy oil residue and ash of oil refineries in the US and Japan, respectively [BESKROVNY, 1993].

Oil and gas fields water

Oil is deposited in the depths of the Earth and, when brought to the surface, it is accompanied by other components of the geological environment: associated water and natural gas. Salt is usually present in the produced reservoir fluid, being a source of anthropogenic salinization of soils.

More than 50 chemical elements have been found in natural water, present as ions, undissociated molecules (including gases), and colloids. However, only a few elements permanently occur in natural water composition.

The principal components in natural water are the six ions containing eight elements: three anions — chlorine Cl^-, sulfate SO_4^{2-}, and hydrogen carbonate HCO_3^-; three cations — sodium Na^+, calcium Ca^{2+}, and magnesium Mg^{2+}. Apart from the six main ions, carbonate ion CO_3^{2-}, potassium ion K^+, and iron ions Fe^{2+} and Fe^{3+} are quite common. Other elements occur in negligible amounts and are referred to as micro-components in water composition. Among those, best known are ions of bromine Br^-, iodine I^-, ammonium NH^{3+}, lithium Li^+, and strontium Sr^{2+}.

Water deposited in oil reservoirs has a variety of distinctive properties that are ecologically important. On the vast majority of oil fields, water contained in the oil strata is highly mineralized. This is either salty water (10–50 g/L) or brines (mineralization above 100 g/L). All petroleum waters are chloride waters. Sodium and calcium are the predominant cations. Waters with the predominant calcium cation belong to the category of extra strong brines with a mineralization above 300–400 g/L.

High haloid content is typical for petroleum waters: besides chlorine, they can sometimes have high content of iodine and bromine (in the range of tens/hundreds and thousands of milligrams per liter, correspondingly). Besides, the waters are rich in boron, strontium, barium, and ammonium. Abundant gaseous components of petroleum waters include H_2S, nitrogen, and helium. Produced fluid contains the salts ($NaCl$, $CaCl_2$, $MgCl_2$, $NaHCO_3$ etc.) that enter the well during oil extraction.

Depending on the deposit composition and its development stage, petroleum and water quantity in produced fluid can vary substantially, although gas and water are always present in crude oil. Oil primary treatment in fields involves removing water and dissolved salts and gases from crude oil. This can result in water and gas discharge into the environment with a detrimental effect.

Reservoir water circulation throughout the oil field cycle results in equipment contamination with natural radionuclides involved in the process cycle. Discharge of petroleum slurries and reservoir water with elevated radioactivity results in background radiation increasing to 20–40 µR/h with peaks up to 1,000 µR/h and more. Radionuclide contamination of filter media, oil sludge, tank cleaning residues, and operating equipment in some cases can reach 3,000 and 5,600 µR/h while the general normal background is 8–12 µR/h. High mobility of radionuclides in petroleum waters facilitates their accumulation in the oil field development process cycle. Therefore, radiation protection measures must be elaborated and implemented both during prospecting, and exploration and during petroleum production.

Radionuclides

Radionuclides are among extremely hazardous substances that are lifted to the surface in the course of hydrocarbon deposits development. Besides oil field connection to reservoir water, radionuclides come from other sources as well. The highest level of radiation hazard exists on the developed fields that are located within the boundaries of abnormal radioactivity areas. For example, uranium concentration in oil varies in the range of two orders of magnitude reaching 10–15 mg per 100 g of oil. Uranium is associated with the resinous-asphaltenic oil fraction and occurs in the form of organic metal complex. Carbonaceous substances in general are effective sorbents of radioactive elements; hence, potentially high radioactivity of certain types of oil must be taken into account as far as the environmental aspect is concerned.

Carbonaceous substance forms soluble compounds with the radionuclides that have a high degree of geochemical mobility. They are easily transferred to water solutions and migrate in alkaline, neutral, and acid media. There are major nature hard bitumen accumulations around the world that contain industrial concentrations of uranium (uranium-bitumen fields). Uranium content in the fields, where it is bonded with the carbonaceous substance, ranges from bulk earth values to tens of percent. In the presence of uranium-bitumen formations, besides uranium, the carbonaceous substance can contain other elements, such as Ra, Mo, V, P, Hg, Se, Ni, Co, Ag [Melkov & Sergeyeva, 1990].

Hydrogen sulfide gases

Hydrogen sulfide (H_2S) is a colorless, highly toxic gas. Therefore, its concentration is strictly limited in domestic gas consumption. Only sulfur-free gas may be used for industrial and domestic purposes (< 0.00139%).

The danger of producing gases with hydrogen sulfide content is twofold and results from both hydrogen sulfide toxicity and its explosion hazard. Dissolved hydrogen sulfide is present in oil or water. It occurs in associated gas and can be formed when water bodies and overwatered soils (bog and meadow soils) are polluted by sulfurous oil. High airborne concentrations of hydrogen sulfide (1 mg/L) lead to intoxication and death of animals and humans. In the presence of hydrocarbons, the maximum permissible concentration of H_2S in air is 3 mg/m^3.

In Russia, oil fields with abnormal hydrogen sulfide content are located in the Caspian region. These are Orenburg (1–5% H_2S) and Astrakhan (20–25% H_2S) fields. When oil is produced from deep horizons (more than 4 km), increased reservoir temperatures and pressures cause the hydrogen sulfide concentration to grow, to the extent of sulfurous gases formation risk.

Mercury-containing natural gases and oil

Mercury is a highly dangerous and barely noticeable admixture in natural gas fields. At atmospheric pressure, natural mercury content in natural gas is $1 \cdot 10^{-7}$ g/m^3. It is higher than natural mercury content in the Earth's atmosphere by as much as an order of magnitude. In

28

certain fields, mercury concentration in gas grows by tens to thousands times, making gas a serious environmental hazard. Under reservoir conditions, mercury concentration in gas is much higher. In Groningen, the Netherlands, the largest gas field in Europe, mercury concentration in natural gas exceeded the saturation limit ($1 \cdot 10^{-3}$ g/m^3), resulting in fallout of free metallic mercury in pipes. As revealed by monitoring observations, a sharp increase of mercury concentration in gas occurs in pulses related to variations in the Earth's crust behavior. The highest Hg concentrations are typical of the fields in the most permeable areas of Earth's crust, where longitudinal and lateral deep-seated faults intersect.

Mercury, at high concentrations in oil, poses a serious environmental hazard. Generally, mercury content in oil remains in the range of $4 \cdot 10^{-6}$—10^{-5}%. It is higher in certain samples, for example, in the Carpathian region, up to 10^{-4}% and oil of the Cymric Oil Field, California, up to $2 \cdot 10^{-3}$%. In this field, high concentration even makes it a small-scale side product. The fields are influenced by the same San Andreas deep-seated fault that is responsible for mercury fields in California (New Almaden, New Idria, etc.). Mercury reserves in petroleum deposits may reach very large quantities. As estimated by N.A. Ozerova, based on hydrocarbon reserves and their average mercury content, mercury reserves in the Groningen gas field, Netherlands, amount to around 400 tons, in the Cymric field, USA, around 1,000 tons [OZEROVA, 1986].

CHAPTER 2.
NATURAL HYDROCARBON FLOWS
IN THE BIOSPHERE

2.1. GREATER HYDROCARBON GEOCHEMICAL CYCLE

The interchange of substance, energy, and information with the surrounding Earth spheres is crucial for the biosphere existence. Practically all known chemical elements are involved in this interchange. However, it is engineered by flows of just four elements — hydrogen, carbon, oxygen, and nitrogen (H, C, O, N) in the form of living matter, water, carbon dioxide, and hydrocarbons.

The diversity of natural hydrocarbon flows constitutes the larger geochemical cycle that includes five interconnected blocks of geochemical processes responsible for hydrocarbons formation and their circulation between geospheres (Fig. 2.1).

I.   *Deep Earth hydrocarbon degassing block.* The larger hydrocarbon geochemical cycle originates in the deep degassing block that uses the space carbon reserve accumulated on the Earth during its formation as a planet. Deep degassing is a natural process of Earth evolution that consists in moving light substance from the core and mantle to upper geospheres, including the biosphere. Besides other chemical elements, deep degassing products always include carbon and hydrogen needed for hydrocarbon synthesis, and elementary natural gases. Thermodynamic calculations indicate that liquid hydrocarbons may also exist in these geospheres.

II. *Lithosphere block.* The lithosphere is the main hydrocarbon-producing reactor on the Earth and a huge reservoir for their accumulation. Carbon enters the lithosphere from a variety of sources: from the Earth's mantle in the magmatic and volcanic process (M) and from fossilized residues of dead organisms buried in bottom sediments of oceans, seas, and lakes (S). Hydrocarbons are formed when exposed to deep heat, pressures, and catalysts in the rock formation processes — lithogenesis and petrogenesis (LP). Hydrocarbons are caught, transferred, and thickened by hot gas and hydrothermal solutions (GH).

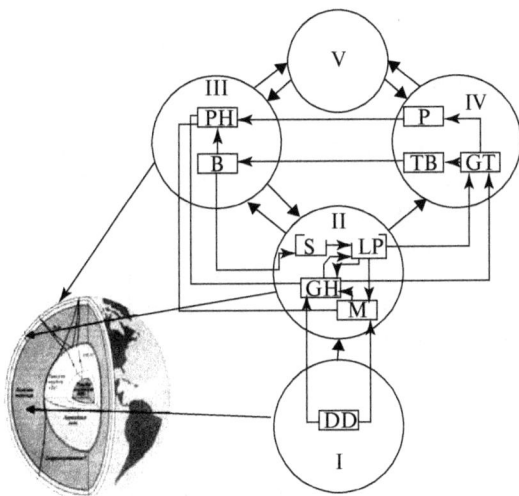

Fig. 2.1. Greater hydrocarbon geochemical cycle diagram [Pikovskiy, 2016].
I — deep Earth degassing block (DD); II — lithosphere block; processes in the
lithosphere block: M — magmatic (volcanic), GH — gas and hydrothermal,
LP — lithogenesis and petrogenesis, S — sedimentogenesis;
III — biogeochemical block, processes: PH — photochemical,
B — biogeochemical; IV — technospheric (anthropospheric) block, processes:
P — production, GT — geotechnical, TB — techno-biogeochemical;
V — space block of hydrocarbon geochemical cycle.

III. *Biosphere block.* This block includes biogeochemical hydrocarbon synthesis by living matter and hydrocarbons recycling and decomposition by microorganisms (B). The hydrocarbons that enter the biosphere from different sources undergo photochemical oxidation (PH) at the Earth's surface and in the atmosphere. Hydrocarbons coming from the lithosphere and technosphere (anthroposphere) are also involved in the geochemical processes of the biosphere.

IV. *Technospheric (anthropospheric) block.* This block is responsible for forced extraction of hydrocarbons from the lithosphere to the Earth's surface, petroleum products transportation, storage, processing, production, and consumption, as well as using them to produce energy and materials (P). Interaction between anthropogenic hydrocarbons and the environment, transformation, oxidation, biological purification of polluted land and water take place in the geotechnical (GT) and techno-biogeochemical (TB) process.

31

V.   *Space block.* Carbon, hydrogen, carbonaceous substances, including hydrocarbons, are common in space. A substantial part of big planet atmospheres (Jupiter, Saturn, Uranus, and Neptune) is made up by methane. Numerous lakes and rivers filled with solidified methane and ethane have been discovered on Saturn's largest moon, Titan. Bituminous substances, polycyclic aromatic hydrocarbons (PAHs), and amorphous carbon have been found on various types of the meteorites that had hit the Earth. Space hydrocarbons were included into the Earth compositioduring planet accretion. The exchange between the space substance and the substance of the Earth's atmosphere and surface continued throughout the entire geological history. Carbonaceous substances participating in this exchange are included into the space block of the hydrocarbon cycle. It is hard to calculate the balance of this exchange reliably.

2.2. HYDROCARBON SOURCES AND FLOWS IN THE BIOSPHERE

Hydrocarbons in the Earth's atmosphere

Hydrocarbons, primarily, methane, are continuously present in the gases of the Earth's atmosphere. Long-term observations of the Earth's atmosphere condition indicate that the average background concentration of methane in the Earth's atmosphere is growing by 1% on average annually. Methane is distributed unevenly in the troposphere. Its concentrations vary in space and time depending on its efflux from the Earth's surface.

Total methane quantity in the troposphere is about 7 trillion m^3 or 7 billion tons. There are different estimates of the average methane molecule "life" span in the troposphere. Some sources give a value of 1 to 7 years [Ehhalt, 1974], others put it at 8−12 years [Isidorov, Kondratyev, 2001]. Therefore, annual methane outflow from the troposphere should be between $0.7×10^9$ and $1.4×10^9$ t. Since troposphere methane concentration varies by a very small amount, atmospheric methane loss must be replenished with the same amount of methane from the Earth's surface. Besides methane, the troposphere contains hydrocarbons with carbon atoms number between 2 and 20. Concentrations of these molecules are smaller than methane concentrations by 1−2 orders of magnitude. Hydrocarbon concentrations dramatically vary in time and space. Their quantity in the ground level atmosphere and over continents is much greater. Atmospheric methane homologues include both natural and anthropogenic hydrocarbons.

Large quantities of carbonaceous compounds in the atmosphere occur on aerosols. Aerosol content in the atmosphere varies strongly between tens and hundreds of $\mu g/m^3$. Both normal and branched alkanes, as well as polycyclic aromatic hydrocarbons (PAHs), have been found on aerosols. Of those, about 80% are located on minute particles that are less than 3.3 μm in size [Isidorov, 2001]. Natural PAHs are carried into the atmosphere as a result of volcanic, postvolcanic, fumarole, and mud volcano activity. A substantial part of anthropogenic carbonaceous substances in the atmosphere is produced by industrial facilities and vehicle emissions.

Hydrocarbons in living matter

Annually, the planet living matter synthesizes, consumes, and releases into the environment around 100–300 million tons of hydrocarbons, mostly, of aliphatic structure. Living organisms easily consume and release volatile hydrocarbons (C_1–C_6). Liquid (C_7–C_{16}) and, especially, high molecular weight hydrocarbons (C_{17}–C_{40}), apparently, have a strong bonding with the organism substance and are released into the environment only after its death and decomposition.

Hydrocarbons have been found in all organs of higher plants. Higher alkanes are present in the cuticle of plants to protect them from drying, epiphytes, and small herbivores. Mainly, they are chains with an odd carbon atoms number. Polycyclic aromatic hydrocarbons (PAHs) have been also found in higher plants. There is no consensus as regards the origin of these hydrocarbons. It is indicated that plants (cabbage lettuce, soy, rye, tobacco) contain benz(a)pyrene, perylene, anthanthrene, benz(ghi) perylenes, dibenzoanthracene, coronene, phenanthrene, anthracene, pyrene, fluoranthene, benzanthracene, and chrysene. These hydrocarbons have been found in plants grown under natural conditions. However, PAHs have not been found in plants grown under climate chamber conditions. The total PAHs content in higher plants is $5 \cdot 10^{-7}$–$1 \cdot 10^{-5}\%$ dry weight. It has been established that plants take up these hydrocarbons from the environment. Some researchers believe that certain PAHs are synthesized in plant organisms.

It is common for plants in nature to release gaseous hydrocarbons. A chemical exchange process between plants, called allelopathy, has been observed for a long time. Besides volatile alcohols, aldehydes, organic acids, and ethers, gaseous hydrocarbons have been found in the substances released by plants: methane, ethane, ethylene, propane, butane, their heavier homologues, as well as volatile compounds like benzene. The process of the release of volatile hydrocarbons by plants takes

33

place only in the presence of light, i.e. it is apparently related to photosynthesis [SANADZE, 1961].

Little data are available on hydrocarbons content in animal bodies. It is known that volatile hydrocarbons, including methane, are formed in the process of fermentation in the intestines of ruminants. D. EHHALT [1974] gives the following values of methane production for different animals: from 15 (sheep, goats) to 100–200 g (horses, cattle) per day. According to D. Ehhalt's calculations, animals release 100–200 million tons of light hydrocarbons per year.

Hydrocarbons are present in the composition of marine algae. Algae normally contain a mixture of normal and branched alkanes predominated by C_{15} and C_{17} hydrocarbons. Normal alkanes are part of the wax-like layer acting as algae protector. Total hydrocarbons content in bodies of marine organisms varies between $1 \cdot 10^{-4}$ and $2 \cdot 10^{-2}\%$ wet weight, while it amounts to $5.45 \cdot 10{-2}\%$ in polluted water bodies [MIRONOV, 1985].

Hydrocarbon state of the soils

Hydrocarbons are among non-specific humus components. [GENNADIEV ET AL., 2015]. Light hydrocarbons make up part of soil atmosphere dispersed gases. High molecular weight non-volatile hydrocarbons are present in the composition of soil lipids — a hydrophobic mixture of organic compounds extracted from the soil matter by alcohol-benzene. In peats, hydrocarbons occur in the composition of "peat bitumen" whose properties resemble those of soil lipids. Moreover, soil receives large quantities of allochthonous carbonaceous substances of natural and anthropogenic origin that fall out of the atmosphere, lifted from the lithosphere, or end up on the soil surface in the course of economic activity. Total hydrocarbon reserves in the global soil cover (excluding arctic, sand, stone deserts, glaciers, and snow patches) are approximately 2–3 billion tons (with the average hydrocarbon content of 0.002–0.003%). Annual flow of lipids to the soil is on average 10–12% of their reserves contained in the soil. Therefore, global inflow of hydrocarbons to the soil could be estimated at 200–300 million tons per year, which is equal to the amount of hydrocarbons annually released by living matter. According to some estimates, PAHs constitute 0.1–0.5 million tons of those hydrocarbons.

To describe the entire complex of hydrocarbon compounds in soils, including both natural and anthropogenic hydrocarbon components, oil and petroleum products among them, the notion of *"hydrocarbon state of the soils"* has been introduced. This term relates to the ratio and composition of natural gases, bitumoids, and individual hydrocarbon compounds (polycyclic aromatic hydrocarbons, n-alkanes, etc.) [PIKOVSKY ET AL., 2008]. Four main hydrocarbon states of the soils have been identified, depending on the nature of the geochemical processes that have formed the soil hydrocarbon complex. They are biogeochemical, emanation, atmo-sedimentation, and injection processes. All hydrocarbon state types can be both natural and anthropogenic in origin.

Biogeochemical formation of natural hydrocarbons in soils is mainly connected with microorganisms' activity. Methane-synthesizing bacteria, called methanogens, belong to Archaebacteria. Methanogens that dwell exclusively under anaerobic conditions have been extracted from water, bottom sediments, bogs, seas, soils, subsurface water, household effluents, and rumen of ruminant animals. The temperature range of 30–40 °C is most favorable for growth of most methanogens. However, there are some thermophilic representatives found in a variety of oceanic and terrestrial thermal springs. These microbes demonstrate the fastest growth rate at 65–70 °C. Optimum pH values for most species are 6.5–7.5. These microorganisms are a vital component of organic matter anaerobic decomposition. Nearly all presently known groups of microorganisms are capable of hydrocarbon synthesis — bacteria, fungi, algae [DEDYUKHINA, ET AL., 1980]. Hydrocarbon content in microorganisms themselves is 0.02–2.69% for bacteria, 0.06–0.70% for fungi, and 0.03–2.88% for algae.

Gaseous hydrocarbons in soils are formed under reducing conditions, and at particularly high rates in bogged and flooded soils (rice paddies). Thus, methane content in bog soils can reach 60% of total soil gases. Besides methane, other light natural gases can be also synthesized — ethane, propane, ethylene, etc. [ORLOV, 1985; ORLOV ET AL., 1986].

Simultaneously with hydrocarbon synthesis by microorganisms, processes of their oxidation by methane-oxidizing bacteria take place in soils. The quantitative aspect of this process has not been sufficiently studied yet. S.S. BELYAEV [1988] established that in the silts of Chyorny Kuchier Lake, around 40% of methane generated by microorganisms was oxidized. According to other researchers, 20–30% of monthly methane production is oxidized by microorganisms. In general, one can assume that the pool of biogenic hydrocarbons exchange between soils and the atmosphere is 150–350 million tons per year, and 200–400 million tons per year if heavy hydrocarbons are included.

Hydrocarbons in ocean bed sediments

Hydrocarbons of different origin are accumulated and transformed in the World Ocean. Hydrocarbons that circulate in the sea environment mainly come from the following sources:

- hydrocarbons contained in the bodies of plankton and benthos organisms;
- hydrocarbons formed from residues of living organisms in the process of bed sediments diagenesis in anaerobic conditions;
- hydrocarbons carried to coastal waters by shore runoff;
- hydrocarbon emanations from deep geospheres or lithosphere petroleum deposits;
- anthropogenic hydrocarbon pollution of the marine environment related to petroleum production and transportation, vessels waste effluent, and other human-made sources.

Ocean and sea coastal areas are high biological productivity locations, where the largest part of autochthonous biogenic compounds, including hydrocarbons, is formed. These areas are also the most exposed to man-induced impact, as pollution mainly happens in ports, at crude loading terminals, and on coastal urbanized territories.

Carbonaceous substances composition in the top layer of present-day marine sediments depends on the climatic zone. According to I.A. Nemirovskaya [2004], dead organic residues in marine sediments of arctic climate are poorly modified. They have higher content of easily hydrolyzable substances, including carbohydrates (24–40%) and humic acids (16–22%). Bottom sediments enriched in organic matter may contain up hydrocarbons in amounts of up to 100 μg/g. Concentrations above those levels are normally related to anthropogenic inflows. Biogenic sources are primarily responsible for the level of aliphatic hydrocarbons content. In shallow semi-enclosed water, elevated hydrocarbons content in bottom sediments may result from high plankton productivity at shallow depths, substantial inflows of terrigenous organic matter and high sedimentation rates. Therefore, shallow water maximum of aliphatic hydrocarbons concentration (>100 μg/g dry weight) caused by natural processes, is confined to relatively thin sediments of gulfs, lagoons, bays, intracontinental seas, and the upper part of open ocean shelf.

2.3. LITHOSPHERE — NATURAL MEGA RESERVOIR FOR OIL AND NATURAL GAS

Global petroleum resources involved in petroleum production are accumulated in the lithosphere, the Earth's solid shell that includes Earth's crust and part of the upper mantle. More than 40,000 oil and gas fields have been discovered in the lithosphere upper layers at depths between hundreds of meters and 10 kilometers. Global proven oil reserves amount to nearly 200 billion tons, while gas reserves amount to

200 trillion m³. A substantial part of these accounted reserves has been already exhausted.

Although oil and gas fields are a global phenomenon on the Earth, their reserves are extremely unevenly distributed in terms of quantity. Around half of all recoverable oil reserves discovered by the beginning of the 21st century are accumulated in giant and supergiant fields that make up less than 0.1% of the known fields number. A third of all world petroleum production is concentrated in the Near and Middle East countries, while a third of global gas production takes place in Russia and the USA. Approximately 40% of world petroleum resources are located on sea shelves and continental slopes.

More than 95% of discovered oil and natural gas deposits are situated in petroleum bearing basins — depressions in the modern Earth's crust pattern varying in structure and size, and filled by sedimentary rocks. Hydrocarbon accumulations occur in rocks of any composition and age (Pre-Cambrian, Paleozoic, Mesozoic, and Cenozoic, up to the Quaternary). The largest number of petroleum deposits has been found at a depth of 1–3 km from the Earth's surface. Dozens of industrial deposits have been discovered below the sedimentation layer, in basin foundations — in the upper part of consolidated Earth's crust. More than 600 petroleum bearing and potential petroleum bearing basins exist worldwide. Groups of fields that are located in close proximity within a basin are combined into oil-producing regions, and closely located regions are united into oil-producing areas.

Hydrocarbon deposits require a natural reservoir that can accumulate and store the liquid or gaseous fluid. Two types of natural reservoirs that are used for industrial petroleum production exist: conventional and unconventional ones.

A reservoir of the first type (*conventional*) is a stratum or lengthy rock mass with free pore and fracture space limited by lithological or hydrodynamic barriers. The top lithological barrier of a natural reservoir is called a caprock, while the bottom barrier is called a fluid seal. Brine water is often deposited on the lithological fluid seal supporting the hydrocarbon accumulation base (hydrodynamic barrier). Conventional natural reservoirs are formed in a broad variety of rocks: strata of permeable porous sandstone, carbonaceous, igneous, or intrusive magmatic rock masses. They include a permeable space formed by interconnected pores and open fractures. The part of a natural reservoir where petroleum fluid is "locked" by lithological and hydrodynamic barriers that prevent it from further movement along the stratum is called a trap. Today, traps where hydrocarbon accumulations are formed exist in a multitude of geological forms and sizes. Oil and gas deposit is the petroleum fluid that filled the trap in a natural reservoir. Trap fluid can occur in different phase states: liquid (oil or oil with dissolved gas), gaseous (free gas or gas with condensate), and two-phase state (the deposit bottom part consists of oil, and the top part consists of gas). Petroleum deposits substantially vary in size. Trap length can amount to tens and hundreds of kilometers. World largest petroleum fluid deposit, the Ghawar field in Saudi Arabia, extends across 200 km. Productive stratum height in the trap varies from 1−3 to 100 meters, and, less commonly, to 200−300 meters.

An aggregate of closely located hydrocarbon deposits confined to one tectonic element is called an oil and gas field. A field can consist of just one or several (sometimes dozens of) deposits (a multilayer field).

The second (*"unconventional"*) type of natural reservoirs is a layer of poorly permeable rocks with a large volume of porous space with the isolated pores that contain oil or gas fluid. Such rocks are usually part of "caprocks" or fluid seals in conventional reservoirs. "Unconventional" reservoirs do not require special lithological or hydrodynamic barriers, since the fluid is not able to move on its own inside of them. Unconventional reservoirs hold enormous accumulations of hydrocarbon resources with billions of tons of reserves (shale oil and shale gas). Hydrocarbons production from unconventional reservoirs requires sophisticated technology of deep rock fracturing ("hydraulic fracturing of formation") to release the free fluid from isolated pores. If the combination of economic and geological conditions is favorable, the size of hydrocarbon resources offsets the economic costs of development. Unconventional objects that are introduced into development to obtain additional hydrocarbon resources are listed below.

- consolidated black clay shales distinguished by strongly hydrophobic properties and high carbon content. Such shales are widely used (especially, in the USA) to produce shale gas (mainly, methane) and shale oil. They can significantly vary in composition and gas content between different fields and even wells. Methane content varies between 79.4 and 95.5%, and ethane content is between 0.1 and 16.1%. Nitrogen and carbon dioxide ratio is up to 9.3% [PIPELINE..., 2011];
- tight poorly permeable sandstones, where substantial gas, heavy oil and asphalt reserves are accumulated across large areas (Canada, USA, Mexico, Venezuela);
- coal-bearing layers in deep horizons that contain enormous quantities of natural gas, methane. In the USA, special-purpose wells are drilled to produce methane from coal-bearing layers.

Artificial withdrawal of huge hydrocarbon quantities from lithosphere accumulations is the groundwork of modern petroleum production.

2.4. NATURAL PETROLEUM OCCURRENCES ON THE EARTH'S SURFACE

Types of natural oil and natural gas occurrences on the Earth's surface

There are several ways and sources of oil and natural gas occurrences on the Earth's surface. Forms of natural petroleum occurrences on the Earth surface are given in Table 2.1, fig. 2.2.

Table 2.1

Types of oil, gas, and bitumen natural occurrences on the Earth's surface

Sources of oil, gas, and bitumen	*Forms of oil, gas, and bitumen occurrences on the Earth's surface*
Natural petroleum migration towards the Earth's surface through cracks and tectonic channels from hydrocarbon accumulations in the lithosphere under water and gas pressure	Petroleum occurrences in the form of individual springs
	Oil and asphalt lakes
	Soils and rocks cemented by oil and bitumen that surround petroleum leaks (brea impregnation) or seepage from bitumen-containing rocks exposed by erosion (incrustations)
	Rock layers that crop out to daylight surface or are deposited near the surface, impregnated with asphalt and heavy oil
	Outcropping veins of nature hard bitumens (asphaltite and kerite) in cavi-ties of faults and factures
	Sudden petroleum outbursts to the surface under high pressure (gas gryphons)
Diffusion and effusion dispersion of petroleum fluid accumulations and flows	Formation of geochemical anomalies in soils and bottom sediments above oil and gas fields
"Cold seeps" are ambient temperature petroleum occurrences that are related to migration via Earth's crust faults from deep lithosphere zones	Mud volcanoes
	Hydrocarbon discharges at deep-water continental margins (gas flares)
	Formation of gas-hydrate deposits in bottom sediments
	Petroleum occurrences in continent rift zones
	Petroleum occurrences in igneous rock masses, mine openings, and circular astroblemes ("impact craters")
"Hydrothermal vents" — oil and gas carried to the Earth's surface by hydrothermal solutions in rift zones and volcanic areas	Black and white "smokers" — hydrocarbon leaks carried with hot water solutions located on sea and ocean beds
	Petroleum occurrences on continents in hydrothermally active re-gions of active volcanic areas

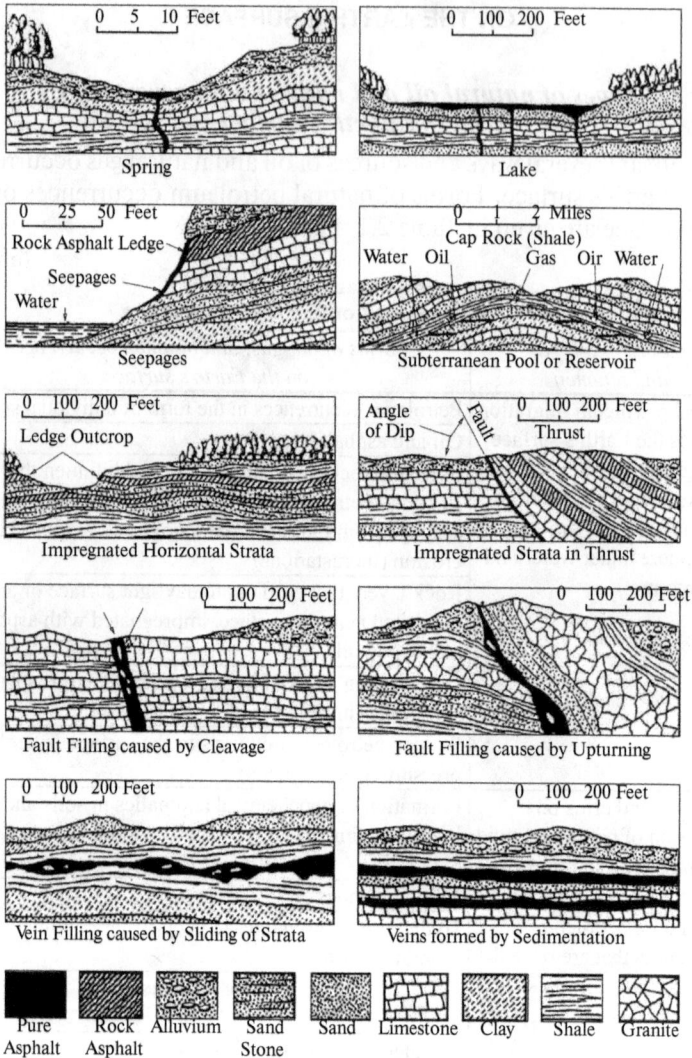

| 0 5 10 Feet | 0 100 200 Feet |
| Spring | Lake |

Rock Asphalt Ledge
Seepages
Water
0 25 50 Feet
Seepages

0 1 2 Miles
Cap Rock (Shale)
Water Oil Gas Oir Water
Subterranean Pool or Reservoir

0 100 200 Feet
Ledge Outcrop
Impregnated Horizontal Strata

Angle of Dip
Fault
0 100 200 Feet
Thrust
Impregnated Strata in Thrust

0 100 200 Feet
Fault Filling caused by Cleavage

0 100 200 Feet
Fault Filling caused by Upturning

0 100 200 Feet
Vein Filling caused by Sliding of Strata

0 100 200 Feet
Veins formed by Sedimentation

Pure Asphalt | Rock Asphalt | Alluvium | Sand Stone | Sand | Limestone | Clay | Shale | Granite

Fig. 2.2. Forms of natural petroleum bitumen occurrences on the Earth surface (BY H. ABRAHAM, 1945).

Oil springs. Oil springs are common across all continents, especially in mountainous and piedmont terrains. Such springs are the most numerous forms of petroleum surface occurrences. Normally, they are small single-point channels that look like springs of water with gas bubbles and floating oil. If oil leakage occurs on a level, it impregnates the soil around the spring within a few dozens of meters in diameter. The size of oil occurrence area remains almost unchanged in time, although the spring may be active for hundreds of years. The rates of outflow from oil springs vary in a very broad range. Normally, they are insignificant. However, sometimes they amount to hundreds of liters per day. For example, according to calculations of K.B. Ashirov, Samarskaya Luka water springs that contain oil films carry up to a total of 0.5−0.6 tons of oil per day, or around 200 thousand tons in 1,000 years.

When oil brought to the surface in the form of springs is no longer made up by fresh portions of movable oil and gas, it is gradually stripped of light fractions and transformed into viscous and hard bitumen varieties — asphalts and kirs. Rocks surrounding petroleum occurrences on the daylight surface, cemented by thickened oil, are present-day and ancient brea impregnations.

Oil occurrences in petroleum-bearing regions are reported both on land and on the sea floor. Such offshore occurrences are known along the California coastline, in the Gulf of Mexico, in the Caspian Sea near Apsheron Peninsula, and in many other locations.

In the early times of petroleum exploration, oil and gas occurrences served as a direct search indicator. However, it was gradually found that natural oil and gas showings normally occurred in areas where the deposits have been destroyed or depleted. On the contrary, oil surface occurrences are normally absent in the areas with large oil and gas fields.

Oil and asphalt lakes are formed by seepage from a compact group of springs (see Fig. 2.2). Water and gas pressure from below forces liquid or semi-liquid oil upwards via fractures and faults, and it is accumulated in depressed landforms. Surface oil layer (between a few centimeters and one meter) solidifies due to atmospheric contact and protects the underlying mass from rapid oxidation. Oil lakes of various sizes are located in petroleum-bearing basins of many countries worldwide — in North and South America, Europe, Asia, and Africa. In Russia, such lakes exist in the northeast of Sakhalin Island. In Azerbaijan, as much as 14 ha of Artyom Island area are covered by asphalt. Trinidad Pitch Lake (Asphalt Lake), one of the world largest oil lakes on Trinidad Island in South America, is well known [LINK. 1952].

Oil sands and shales. Enormous reserves of thickened and sorbed oil are accumulated in the oil sands and shales that occupy large areas near the Earth's surface. Once, these rocks were impregnated with petroleum. However, after this terrain was raised and erosion processes began, these layers became exposed to atmospheric contact and lost a substantial part of gas and light hydrocarbon fractions. The remaining high-molecular-weight components have completely lost their mobility and developed a strong bond with the surrounding rocks. These resources are developed using the open-pit method, in quarries. Often, these strata reach the daylight surface. Oil that is contained in them produces incrustations on rocks surface.

The largest oil sand deposits that contain 8–12% of hard bitumen are situated in Western Canada, in the state of Alberta. Bituminous sands area in this region exceeds 70,000 km². Hard bitumen inclusions in rocks occur from the surface level to the depth of 600 meters. Nature hard bitumen geological reserves amount to around 400 billion tons. Half of them are found in the Lower Cretaceous oil sands, the other half is confined to the underlying strata of Paleozoic carbonates [ABRAHAM, 1945].

Large areas of rocks saturated by hard bitumen and deposited near the Earth's surface are known to exist across different continents. These include the Orinoco Belt in Venezuela with estimated reserves of over 600 billion tons of heavy crude and bitumen, and large territories in the Volga–Ural region and on the Siberian platform.

Gas gryphons. Natural hydrocarbon gas shows are more common than natural oil shows. This is explained not only by the fact that gases have a better migration capacity but also because not all subsurface natural gas accumulations are directly linked to oil deposits. Normally, gas leakage from the depths occurs spontaneously and is hard to observe. In contrast to oil, gas is not capable of forming even temporary accumulations at the Earth's surface. However, a phenomenon exists of sudden gas and water discharges, when those occur under a super-high subsurface pressure. Those discharges are rare but always result in disastrous consequences.

When drilling deep in regions with super-high formation pressures, aggressive gas and water can penetrate through the strata and rise to the surface in the form of water and gas gryphons. Where such leaks occur, rocks acquire flowing properties. As a result, 40-meters high drilling rigs along with all of equipment and surrounding facilities sink under the ground, and are replaced by a lake, whose water bubbles with outflowing gas for many years to come. Such incidents were recorded in the Caspian (the Berikey field in Dagestan and the Oil Rocks field in Azerbaijan), on the Kunzhinsk field in the mouth of the Pechora River and other places.

Gryphons and explosions can also occur naturally, as a phenomenon unprovoked by man.

Cold seeps

Cold seeps are a variety of petroleum shows in the Earth's surface with ambient temperature. They can be only partially related to hypothetical oil and gas fields in the upper part of the lithosphere. These shows are common and even global. They are caused by internal geospheres dynamics and degassing in the process of Earth evolution. These petroleum shows are controlled by large Earth's crust tectonic elements.

Cold petroleum occurrences or discharges to the Earth's surface are connected with two types of sources. The first type is concentrated discharges via mud volcano channels or active areas of large faults (in rift zones) of the lithosphere. The second type is a dispersed gas flow that becomes stronger in the areas of present-day tectonic activity. The activity of this type of flows results in formation of extensive gas-hydrate fields on ocean beds, and in continuous replenishment of methane reserve in the Earth's atmosphere. On their way, these flows could possible saturate groundwater and rock layers with gas, transforming them into "unconventional" sources of crude hydrocarbons.

Mud volcanoes. Mud volcanoes are formed when gases (mostly hydrocarbon in their composition) and reservoir waters with softened silt mass or hard rock debris are periodically carried up via localized radial channels and discharged to the land surface or ocean bed. Mud volcano's eruptive part is essentially cones of various sizes, mounds, or salses built from eruption products. Mud volcanoes are quite common. The majority of volcanoes are confined to the Alpine-Himalayan, Pacific, and Central Asian mobile belts of the Earth. Around 1,000 mud volcanoes exist on the Earth. Around half of them are concentrated in the South Caspian area, and most of them are located in the territory of Azerbaijan.

Some giant mud volcanoes (that are especially common in the territory of Azerbaijan) are 400–450 meters tall, with a crater platform area of 900–1,000 m^2, while the total solid discharge volume at the time of eruption exceeds 2,400 million m^3. Most craters are between a few meters and a few dozen meters in diameter. They are concentrated along fracture faults or anticlinal fold axes.

The channels through which the matter escapes to the surface go down very deep and are essentially sections of Earth's crust faults. Judging from the debris composition, volcano origins are located in the sedimentary mass, at a depth of over 7 km. For the origin to be able to discharge a mass of material to the surface, it must periodically create an abnormally high water and gas pressure that exceeds the lithostatic pressure. This pressure, as ascertained BY P.N. KROPOTKIN AND B.M. VALYAEV [1981], can be only ensured by fluid forced into the origin from deep geospheres.

43

Therefore, the water and gas discharged to the Earth's surface during eruption are most probably a mixture of deep-seated and sedimentary matter from large depths.

The gas of mud volcanoes in the Caucasus, Turkmenistan, and Sakhalin, as a rule, mainly consists of methane. Nitrogen and heavy natural gas quantities are very low, while the inert argon, xenon, and krypton are measured in fractions of percent. Gases of just a few volcanoes in the Kerch region and on Sakhalin Island contain carbon dioxide besides methane. It is not unusual for mud volcano emissions to contain oil. This was the reason why mud volcanoes were thought to be a direct indicator of subsurface oil resources at the stage of early studies. However, it was found that mud volcanoes could as well have no relation to oil or gas fields. Gryphons that occur during deep exploratory wells drilling could also have a mud volcanic nature and result in catastrophic accidents.

Underwater mud volcanoes are common in the Caspian. More than 160 volcanoes have been discovered in the South Caspian water area. Mud volcanoes release gas, low-mineralized water (often with oil films), and volcano sludge to the water surface. Mud volcanoes in the sea appear as separate islands, banks, and uplands. Most mud volcanoes in the sea are located at shallow depths (up to 100 m). Eruption of island and shallow-water (up to 20 meters deep) volcanoes often involves ignition of released gases. The flame can reach 150 m in height. Eruptions of sea mud volcanoes are a serious hazard for sea and island constructions and underwater pipelines.

Based on the data of mud volcano gas emissions just in the Azerbaijan land territory, the total gas quantity released by sea "flares" of the South-Caspian basin alone amounts to tens of trillions cubic meters.

In discharge areas of deep-water cold hydrocarbon seeps, unusual outbreaks of living activity are observed for chemosynthetic benthos and plankton communities, whose food sources include hydrocarbons.

Gas hydrates on ocean beds. Gas-hydrate accumulations are quite common, but all of them are confined to fluid discharge areas at deep-water continental margins in deep-seated fault zones. These accumulations are located near the seabed or at shallow depths in their stable zone. Hydrate accumulations can be related to discharge of both concentrated and dispersed fluid streams on seabeds. International estimates indicate that potential hydrated gas resources amount to $1.5 \cdot 10^{16}$ m^3, while explored natural gas field reserves are $150 \cdot 10^{12}$ m^3 [KVENVOLDEN, 1993]. However, no commercial technology is yet available to recover and use this type of resources.

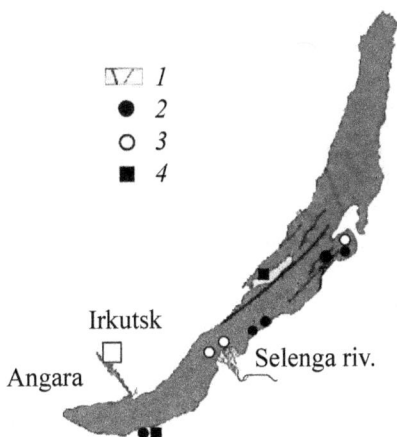

Fig. 2.3. Natural oil and natural gas shows in Baikal Lake
[KHLYSTOV ET AL., 2007, REVISED AND AMENDED SIZYKH ET AL., 2004]:
1 — faults; *2* — oil and bitumen; *3* — gas; *4* — coal.

Oil and gas showings in the Earth intracontinental rift zones. Intracontinental rift zones of the Earth are narrow areas of the Earth's crust tension characterized by high tectonic and seismic activity. These are linear, elongated, and rather deep trough-like depressions (grabens) bordered by large deep-seated faults. Continental rifts along with oceanic rifts are included into the global rift system of the Earth. Some rifts are hundreds of kilometers long, and a few dozens of kilometers wide.

Surface petroleum showings in the rift zones are thousands of years old and are continuously renewed. Major oil and gas fields (in Western Siberia, the North Sea, etc.) are related to the ancient intracontinental rift zones.

Baikal Lake depression could serve as an example of a modern intracontinental rift zone where petroleum showings occur in the Earth's surface.

Oil showings in Baikal Lake were found as early as in the 18th century. Multiple attempts of oil exploration have been made on the coast of Baikal, but none of them was successful. Today, oil and gas springs occur on the surface everywhere along the lake coastline (see Fig. 2.3). They are especially numerous on its eastern coast. Gas bubbles with oil drops are continuously carried up through tectonic faults making up small individual springs. Sometimes, the fluid flow intensifies in certain locations, and small discharges of oil and gas mixture occur. Hydrocarbons are dispersed in the mass of water, partially saturate the sediment, or are thrown on the shore. In winter, the layer of ice acts as a barrier, and hydrocarbons concentrate beneath it.

45

Calculations indicate that all petroleum occurrences of Baikal Lake combined produce at least 15 tons of oil per year, while the total flow rate of gas leaks in the western part of the Selenga River Delta reaches 10,000 m^3/day. However, although new channels of hydrocarbon leakage keep appearing, their total mass and distribution area remain the same. Most natural gas consists of about 90% of methane and more. Its homologues (ethane, propane, butane, isobutane) make about 1%. In certain areas, gas composition is characterized by higher nitrogen (2 to 20%) and carbon dioxide (up to 2%) content. Oil is heavy with a specific weight of 0.965 and contains 1.5% of paraffin and 0.3% of sulfur. Many aromatics, including polycyclic hydrocarbons, are observed in the oil composition.

The distinctive feature of Baikal Lake consists in presence of all types of cold seep hydrocarbon shows. They include oil and gas occurrences in the form of individual springs, deep-water fluid discharges in the form of gas flares, mud volcanoes, and gas-hydrate fields in the bottom sediments near the lake floor. At the same time, all of these hydrocarbon flows have no negative impact on the unique purity of the lake water. Only anthropogenic activities have significant negative impacts on the ecosystems of Baikal Lake.

Hydrothermal vents

Hydrothermal vents are occurrences of hydrocarbon-containing hot springs (hydrothermal solutions) on the surface of continents and ocean beds. This phenomenon is quite common. Such springs are especially numerous in continental areas of modern volcanic activity. Their origin is related to post-volcanic processes.

Petroleum occurrences in hot hydrothermal solutions on the ocean floor (black and white "smokers"). Intensive thermal water discharge into the World Ocean occurs in the rift zones of mid-oceanic ridges and island arc areas. Occurrences of these solutions are accompanied with a black or white plume rising hundreds of meters above them. Because of the plume color, underwater hydrothermal occurrences have been called black or white smokers. The plume of black smokers mainly consists of finely dispersed sulfide material, while with white smokers it predominantly consists of amorphous silica [BUTUZOVA, 2003].

When superheated hydrotherms are cooled, they release hydrogen, helium, chlorine, hydrogen sulfide, hydrogen chloride, carbon oxides, and hydrocarbons. Methane is always present, often in large quantities. At the same time, in many places the superheated solutions cool on the seabed and condense into oil that saturates the sulfide structures surrounding the leaks and spreads across the seabed penetrating into the bottom sediments. Fig. 2.5 illustrates the ocean and continental areas where oil and PAHs separated from hydrothermal solutions were studied.

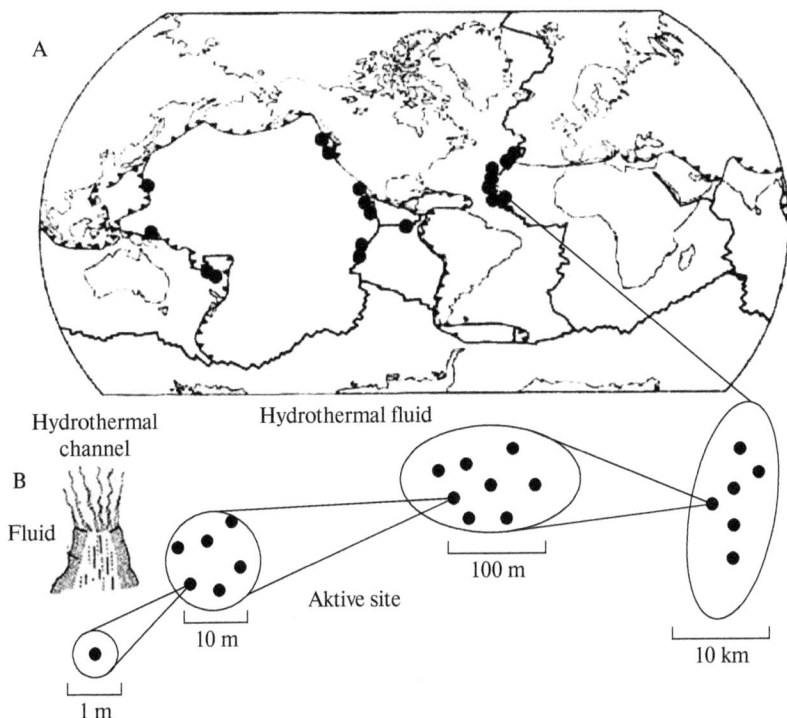

Fig. 2.4. A — Active hydrothermal fields on the ocean floor
(Galkin, 2002, p. 44); B — Channel opening whereby the hydrothermal fluid
[Haymon, Macdonal, 1985, p. 447].

Hydrothermal oil composition and appearance are, in general, similar to the oil of petroleum-bearing basins. Its distinctive difference from regular oil is the fact that it is rich in aromatic structures, in particular, polycyclic aromatic hydrocarbons (PAHs). Oil composition may be substantially different, depending on the area. Both substituted structures (homologues of naphthalene, phenanthrene, benzofluorene, chrysene, pyrene, benzo[a]pyrene, perylene) and unsubstituted structures (pyrene, tetraphene, benzo[a]pyrene, perylene, benzo[ghi]perylcne, anthanthrene, coronene) are identified in the PAH composition.

Hydrothermal solutions on ocean and seabeds leak through individual single-point channels in the form of carbonate-silico-sulfide build-ups of several meters in height. Such channels are grouped into fields of up to 100 meters in size, and the fields are grouped into regions with the size of 10 km and more (see Fig. 2.4).

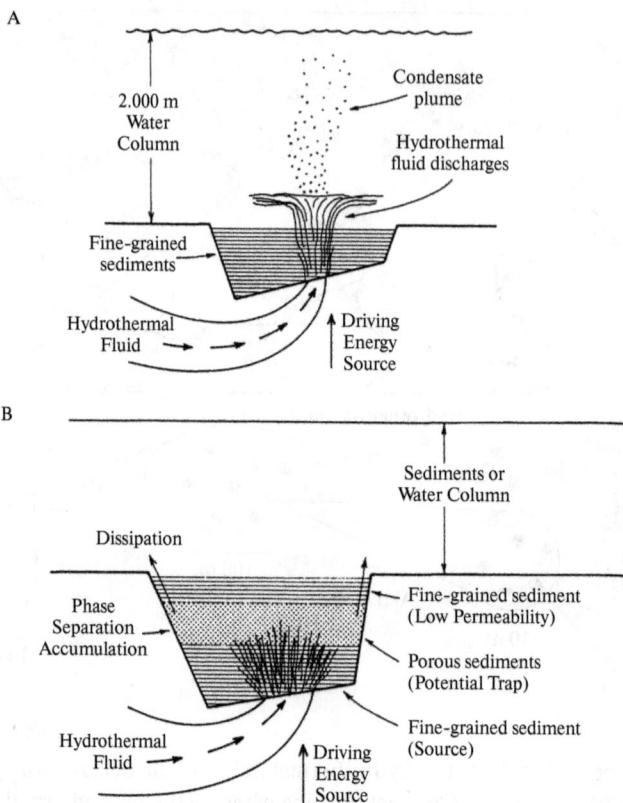

Fig. 2.5. Schematic models of hydrothermal oil shows in hot seeps
[DIDYK AND SIMONEIT, 1989; SIMONEIT, 1995].
A — open system, fluid discharged into water mass.
B — Fluid accumulated in sedimentary rocks or bottom sediments.

Rich and unusual hydrothermal biotic communities have formed around the buildups produced by "smokers". This happens in water saturated with hydrogen sulfide, heavy metals, and hydrocarbons. The temperature of water released by black smokers can reach 300 °C and more. Uniqueness of hydrothermal communities consists in the fact that large heterotrophic animals live off chemoautotroph bacteria that use the compounds reduced in the process of chemosynthesis, instead of photoautotrophs. Bacteria cover the floor surface and "black smokers" buildups in a complete layer that forms bacterial mats. Special methanophilic fauna is concentrated around the springs [GALKIN, 2002].

Hydrothermal fluid occurrences on the ocean floor result in condensation of the hydrocarbons that they contain when the temperature decreases.

This condensation can occur at the time of fluid discharge into water mass (see Fig. 2.5 A) or into porous bottom sediments (see Fig. 2.5 B). In both cases, hydrocarbon anomalies are formed in the sediment or water near these occurrences — gaseous, liquid, or solid hydrocarbons [Simoneit, 1995].

Petroleum occurrences in continental areas of volcanic and hydrothermal activity. Natural gases, especially methane, are frequently encountered components of present day continental hydrothermal springs. In areas of their occurrence, oil drops and films condense out of hydrothermal fluid vapor with the decrease in solution temperature. Oil and natural gas shows have been identified in Uzon caldera, Kamchatka, Yellowstone National Park, USA, Wai-O-Tapu Geothermal Area, New Zealand, and Lake Tanganyika, Tanzania. As with deep-water springs, continental hydrothermal oil greatly varies in its composition even within a single area. Light fractions of Uzon caldera oil contain aliphatic and aromatic structures (with the former being predominant). In heavy oil fractions, naphthenic hydrocarbons play an important role. The composition PAHs in Uzon oil is similar to that of oceanic thermal springs.

PART TWO
PETROLEUM ANTHROPIZATION

CHAPTER 3
ENVIRONMENTAL IMPACT
OF THE PETROLEUM INDUSTRY

3.1. GENERAL DESCRIPTION OF PETROLEUM ANTHROPIZATION

Petroleum anthropization is the environmental impact and its consequences produced by oil and natural gas exploration, extraction, transportation, storage and processing industries, as well as the economic activities that involve products of such processing. Adverse impacts of petroleum industries on the Environment are a global phenomenon.

Although oil production and transportation technology is continuously improving, pollution remains a pressing problem. Massive oil spills on land and in the sea, despite being relatively rare, inflicts a lot of damage, and hurts ecosystems that are far from the initial pollution location. Main potential sources of anthropogenic hydrocarbon flows are listed in Table 3.1.

Table 3.1
Main petroleum production systems that pollute the environment

Process systems	*Environment pollutants*
Petroleum production — oil and gas fields	Crude oil, saleable oil, condensate, mineralized reservoir water, drilling mud, chemical agents, soot left from gas and condensate combustion
Oil and petroleum products transportation by railway and by water	Oil and petroleum products
Main crude oil and gas pipelines	Saleable oil, liquid petroleum products
Oil refineries	Natural gases, carbon, sulfur, nitrogen oxides, petroleum products, and their components
	Hydrogen sulfide, carbon and sulfur oxides, phenols, PAH, and other flare hydrocarbons
	Liquid fuel, residual fuel oil
	Effluent water at treatment plants
Petroleum and gas storage	Oil, petroleum products, natural gases

3.2. IMPACTS OF OIL AND GAS PRODUCTION ON THE ENVIRONMENT

Tables 3.2 and 3.3 list the main impacts on the environment and the way it changes when exposed to petroleum production.

Table 3.2

Environmental changes and their main factors in oil fields

Factors	Type of change
Types and intensity of construction work (construction of industrial facilities, utility networks, and amenity buildings)	Mechanical damage of the land surface and natural vegetation cover
Types and number of vehicles used in the field, intensity of their op-eration	Surface mechanical damage, accumulation of the toxic products that are pre-sent in vehicle exhaust
Composition and quantity of spilled oil	Geochemical changes of soils, surface and groundwater; disruption of normal plant growth and deg-radation of vegetation
Composition and quantity of spilled field wastewater, including oil reservoir water	
Composition and quantity of spilled drilling water and agents	
Composition and quantity of natural gas released to the atmosphere and burnt in the atmosphere and flares	Air, soil, water, and vegetation pollution, thermal impact on vegetation

The substances that are primarily responsible for anthropogenic flows on an oil and gas field are production waste from the drilling process, well maintenance, extracted products collection, and primary petroleum treatment. Field waste that produces the anthropogenic flows on the field and surrounding territory includes:

- oil, gas, and wastewater obtained by produced fluid separation and petroleum primary treatment, underground water directly from production strata;
- oil reservoir edge water, including edge water used to maintain the formation pressure;
- gas from oil deposit "gas caps";
- products of gas combustion in flares;
- drilling mud used to flush boreholes while drilling;
- chemical agents used for hole treatment to increase production rates: various additives, corrosion and scale buildup inhibitors, petroleum pro-ducts.

These substances are released into the environment for a variety of reasons: technology faults or violations, equipment poor quality, wear or deficiency, as a result of emergencies, reservoir drive variations, etc.

51

Table 3.3
Groups of anthropogenic geochemical impacts on the environment at various oil production stages (according TO SOLNTSEVA, 1998)

Impacts			Groups of environment pollutants and their composition
Types of work	Sources	Causes	
Prospecting and exploration, and oil (gas)-field construction			
Road construction, construction of site bunds, drilling wells	Wells, pits, liquid circulation systems, heavy vehicles	Emergency emissions, poor equipment sealing, effluent water discharge, pit outbreaks and overflowing	Flushing liquids, drilling sludge, petroleum products, cement, gypsum, silicate, soda, lime, salt and other process solutions, corrosion inhibitors, surfactants
Field operation			
Oil production and transportation, pumping water into wells	Sumps, pump stations, production and injection wells, pipelines	Metal corrosion, poor sealing, pipeline failure	Crude and desalted oil, wastewater of various mineralization degree; hydrocarbons, phenols, surfactants, and their additives ($Ca(NO_3)_2$, $Mg(NO_3)_2$, Na_2SiO_2, HCl, KCl), corrosion inhibitors, hydrogen sulfide, salts (mostly chloride), rare and trace elements, pol-ymers, alkali
Oil collection and initial preparation in the field			
Crude oil collection, separation, disposal of associated gas and condensate	Oil tanks, pipelines, flare systems	Light fractions lost to storage, hydrocarbon pyrolysis, pipeline corrosion	Products of associated gas and condensate incomplete combustion (PAH emissions, including benzo[a]pyrene), oil, hydrocarbon oils, and other petroleum products, nitrogen and sulfur compounds (including hydrogen sulfide), surfac-tants, phenols, corrosion inhibitors, salts, rare and trace elements, etc.

Producing well clusters are the most numerous centers responsible for generating anthropogenic substance flows on the field. Generally, all wells of a cluster are constructed simultaneously, that is why a cluster can be viewed as a single source of anthropogenic flows. In the service life of a field, several hundred wells would be normally drilled. Well drilling and treatment, oil production and treatment are performed continuously throughout the entire period of the field existence. The field in general and the surrounding territory are continuously exposed to the impact of the above-mentioned substances, although their relative quantity varies from area to area.

At wells, natural environmental pollution by drilling and production occurs in emergencies, during repair work, by hydrocarbon leakage through flange connections (gaps in seals and valves), pipeline ruptures, oil spills from separators and sumps emptying, as well as due to other well operation faults. In an oil field, a well mainly discharges the produced fluid, which is crude oil containing solution gas and reservoir mineralized water. The amount of water depends on the deposit water-cut. Water and oil ratio in produced fluid varies from 1:100 to 100:1. Produced fluid pollutes soils and water with oil and soluble salts. Oil and reservoir fluid chemical composition varies depending on the region. That is why their environmental impact is different even if all other conditions are the same.

Drilling and field wastewater presents a serious hazard during well drilling and operation. In oil-producing countries worldwide, their volume is growing fast and by far exceeds the volume of produced oil. As a sewage system is not available, field waste is dumped into the nearest water bodies or bogs resulting in heavy pollution of surface and groundwater.

Oil and gas exploration and production quite often lead to uncontrollable flowing of wells. In some cases, oil and gas blowouts can be dealt with rather quickly, while in others, the problem remains unsolved for weeks and even months. Often, the gushing fluid ignites. In such cases, it may be extremely difficult to deal with the consequences of the accident. Irreversible consequences of uncontrollable well flowing are related to formation of gas gryphons — gas flows that penetrate through the rocks. Gryphons lead to rock softening and formation of deep funnels that consume the borehole equipment. Enormous environmental damage of uncontrollable oil and gas blowouts is evident for all environmental compartments (atmosphere, water bodies, soil, minerals, etc.).

To store drilling mud and drilling sludge, special open pits are constructed on drill sites, being a source of environmental pollution. Drilling sludge generated during well drilling can contain up to 8% of oil and up to 15% of organic chemical agents that are used in drilling mud.

Besides drilling operation, primary oil treatment also generates a considerable amount of oil sludge. In this case, sludge can contain up to 80–85% of oil, up to 50% of mechanical impurities, up to 67% of mineral salts, and about 4% of surfactants.

Among drilling fluid components, salty water and water-oil emulsion have the highest migration ability and are capable of creating pollution areas with various sizes and configurations. Soils, ground and surface waters take the bulk of these flows impact.

According to Yu.P. Gattenberger, oil and gas fields affect the geological and natural environment "from the top" (from the surface) and "from below" (from the rock mass). "Top" impact (*anthropization from up*) prevention and mitigation of its consequences are ensured by production standards, equipment sophistication and reliability, compliance with preventive maintenance procedures, complete disposal or burial of effluent water and waste, and monitoring of potential fault causes. Anthropization from up is typical for the oil deposits that are developed by reservoir flooding and for effluent water discharge areas. Its impact on fresh water horizons and the Earth's surface hydrosphere can result in water salinity increase, groundwater level rise, formation of springs and gryphons, swamping, and waterlogging of the territory, The channels connecting deep strata with the surface can be the strata themselves if they enter the surface not far from the field, disjunctive faults and tectonic fractures, unreliable caprock, or untight boreholes. Protection of geological environment from anthropization consequences is ensured by reliable well design, high-quality cementing, raising cement up to the well head, timely elimination of cement leaks, monitoring hardened cement condition in the wellbore during operation, and repair work. "Bottom" impact (*anthropization from down*) is caused by petroleum extraction from the reservoir. This impact originates due to the changes in pressure of the subsurface layers that contain the fluid. Anthropization from down is specific for gas and oil deposits and for groundwater supply points. It can result in surface sagging and depletion of fresh water horizons, as well as area desiccation. To protect the environment, petroleum deposit development must be designed in a way to predict potential land sagging. To prevent sagging, anthropization from down must be constrained.

3.3. THE ENVIRONMENTAL IMPACTS OF OIL, PETROLEUM PRODUCTS, AND NATURAL GAS TRANSPORTATION

Oil and natural gas transportation lines from production sites to processing and consumption locations extend across enormous distances measuring tens of thousands of kilometers on land and across the oceans. Crude hydrocarbon loss and its release into the environment are inevitable on these lines, resulting from both emergency spill and inherent leaks.

Crude hydrocarbon transportation means include pipelines, water, railway, and road transport.

Pipeline transportation. Petroleum is moved across long distances from production areas to oil refineries and other consumers (for example, reloading to sea vessels) through a system of main pipelines (crude oil pipelines, petroleum product pipelines, and gas pipelines).

All main pipelines have the same structure regardless of their intended use: 1) feeding pipelines (from production or processing locations); 2) main transportation facilities; 3) intermediate compression station; 4) linear infrastructure (pipelines); 5) point of destination.

The environmental hazard from pipeline operation consists in oil, gas, and petroleum product spills. Unless quickly contained, small spills can lead to oil substances accumulating in soils and subsoil sediments and to surface and groundwater pollution. In the absence of aeration, gas leaks (for example, in lows) can result in build-up of explosive concentrations and spontaneous explosions. Particularly serious damage results from pipeline ruptures caused by mechanical damage and corrosion of pipes under the impact of humans or natural causes (for example, seismic events). In this case, a single accident can lead to tens and hundreds of tons of oil spilling from the pipe onto soil and water surface, making vast areas unusable. According to expert estimates, pipeline accidents account for almost half of total oil volume that ends up in water bodies.

When environmental hazard is not taken into account during pipeline construction, serious negative changes in the environment could be the result. For example, in tundra and forest-tundra regions, it can lead to changes of microclimate and soil moisture regime and disturb the thermal balance of permafrost. Destruction of vegetation cover sensitive to mechanical impact is observed.

Long-distance petroleum transportation through pipelines has many advantages. However, pipelines cannot replace other means of hydrocarbon crude transportation across relatively short distances or away from pipeline routes, as well as on transoceanic directions. Therefore, the risk of environmental pollution as a result of damage of transportation vehicles or routes exists in practically every corner of the world.

Railway transportation. By railway, oil, liquefied gas, and petroleum products are transported in closed cars and are packed into tanks or special containers. Environmental hazard of railroad transportation stems from catastrophic consequences of accidents that involve train cars derailing or train collision. Spills of transported product and strong impacts result in ignition and powerful explosions causing substantial damage to nearby residential areas and the environment.

Road transportation is mainly used to deliver petroleum products to the consumers that are located at a short distance from petroleum depots. Petroleum products are transported in road tankers of various capacities (less than 2 to 15 tons) designed for different petroleum products, as well as in special containers — metal drums, cans, bottles, and cylinders.

Road transportation, as the volume per vehicle is relatively small, causes much less harm to the environment in case of an accident, than other modes of transportation. However, petroleum products are usually moved on roads that are busy with traffic and pass through populated areas. That is why tank truck accidents can result in higher loss of life than accidents with other oil and petroleum products transportation methods. Besides, soil and water pollution tends to be rather heavy in areas where petroleum products are loaded into road tankers because the number of filling operations per unit of volume is much higher than with other transportation means.

Water transportation is a very common method of oil and petroleum products transportation, particularly, in Russia, where navigable rivers are abundant, and sea coastlines stretch across long distances. Dry-cargo vessels and oil tankers are used for transportation by river and by sea (river and sea vessels, correspondingly).

The global tanker fleet transports by sea more than 1 billion tons of oil annually. Submarine tankers based on large submarines are being designed for operations in the polar seas.

Liquefied natural gas transportation by tankers is a special type of water transportation. Liquefied gas is carried in the special spherical reservoirs that can withstand high pressure. Although the tanker fleet provides many economic benefits, it also poses a great danger for the environment. Tanker accidents occur less often than railroad accidents, but they result in spillage of enormous oil quantities causing grave harm to ecosystems. For example, the Torrey Canyon tanker shipwreck off the shore of England in 1967 led to more than 30,000 tons of oil being spilled into the sea. The accident of Amoco Cadiz supertanker that broke up near the French shore in 1978 resulted in 220,000 tons of oil being spilled into the sea. In 1989, the Exxon Valdez tanker was wrecked off the shore of Alaska spilling 40,000 tons of oil, while the Braer tanker spilled 85,000 tons of oil near the Shetland Islands in 1993. Annually, a total of 300,000 to 400,000 tons of oil is released into the marine environment as a result of discharge from tankers.

3.4. IMPACT OF OIL REFINERIES ON THE ENVIRONMENT

An oil refinery production

Oil refining is a large-scale production based on transformation of oil, its fractions, and petroleum gases into salable petroleum products and feedstock for petrochemical plants, organic, and microbiological synthesis. The production implies a combination of physical and chemical technological processes that are implemented at oil refineries and that include feedstock treatment and its primary and secondary processing.

Oil refineries are localized sources of hydrocarbon anthropogenic flows into the environment.

An oil refinery territory is divided into several areas based on their purpose: (1) processing area with the numerous reactors that produce petroleum products; (2) petroleum tank park — the area for storing petroleum products; (3) area of effluent treatment facilities, settling ponds, and pits; (4) flare system area; and (5) household buildings and plant support facilities area. Each area forms its individual environmental situation that affects the plant in general and the territory that surrounds it.

Processing area. Crude oil received at the refinery is prepared for processing by additional multistage dewatering and desalination. This process is completed using the electrical desalter unit (EDU).

Crude primary processing takes place in tube heaters and rectification columns — vertical cylindrical vessels that separate oil into temperature fractions under atmospheric pressure (AT) or under vacuum (AVT). In AT units, oil is heated to 350–360 °C and fed into the rectification column where oil vapor is condensed and separated into gasoline, kerosene, diesel fuel, and residual fuel oil depending on their stripping temperature. Light fractions are distilled from residual fuel oil at temperatures of 360–450 °C and used to produce various hydrocarbon oils in AVT plants. Heavy distillation residue, tar oil, is supplied to thermal and catalytic oil processing units. Gaseous hydrocarbons that cannot be condensed into the liquid phase are transferred to the gas fractionation unit (GFU).

To obtain petroleum products from tar oil, its viscosity has to be reduced. This is done in the visbreaking unit where the feedstock is processed at temperatures of 440–500 °C and pressures of 1.4–3.5 MPa.

Numerous powerful units in the processing area are designed for heavy oil residue deep conversion to produce light petroleum products, bitumen, coke, and pitch to improve petroleum products quality and purity and to produce petrochemical feedstock.

Deep petroleum conversion is completed in thermal and catalytic processing units. Thermal processing units are used for a variety of thermal cracking processes (producing light petroleum products by splitting large molecules at temperatures of 470–540 °C and pressures of 4–6 MPa); coking (light petroleum products and solid coke production at lower pressures); pyrolysis (unsaturated hydrocarbons production for the petrochemical industry at temperatures of up to 900 °C).

In the catalytic processing units, the following processes are used that involve catalysts: cracking at lower temperatures and pressures, reforming (gasoline fraction processing to increase the octane number and separate aromatic hydrocarbons); hydroprocessing (oil fraction processing with introduction of hydrogen).

The units inside the processing area are linked by a dense network of pipelines running both under and on the surface.

Before dispatching salable products to the consumer, most bulk products obtained in different units, for example, motor fuel produced by straight distillation and by cracking, are mixed, which makes it impossible to separate those units.

Petroleum tank park. A petroleum tank park is designed for storage and shipment of salable finished products — light and dark liquid petroleum products. For this purpose, steel tanks of 400 to 5,000 m3 in capacity are constructed and linked to the processing area by pipelines. All tanks are marked with the corresponding petroleum product (gasoline, residual fuel oil, etc.) Tanks have floating roofs. Their bottoms are seated on sand cushions. Tanks number and capacities are determined based on the product storage period required. At least three commercial tanks are provided for each product. Normally, a commercial tank is designed to hold two weeks reserve. However, in some cases the storage period can be significantly longer depending on the conditions of salable products shipment.

Effluent treatment plant. Effluent treatment plants are designed to treat polluted wastewater generated in the production process. Apart from mechanical and chemical treatment units, water undergoes treatment in a designated area using absorption fields, biological settling ponds, traps, pits, etc.

Flare system. The flare system is constructed to prevent harmful gas and vapor emissions to the atmosphere by incinerating them. The flare receives gases and vapor from production equipment, emergency, and process releases. The flare unit consists of the feeding pipeline, open burning stack, and fire containments.

The hazard of environment pollution

Oil refineries are dangerous for the environment due to the following factors:
- product toxic properties;
- fire and explosion hazard;
- atmospheric pollution;
- pollution of surface and groundwater;
- pollution of soils and ground.

The overwhelming majority of substances used in oil refining and petrochemical industry are fire and explosion hazardous, harmful (toxic), and carcinogenic.

The hazard of environment pollution from petroleum products comes, mainly, from their thermal production methods (cracking, pyrolysis) that result in formation of polycyclic aromatic hydrocarbons. Despite worldwide ongoing work to reduce motor fuel toxic properties, the problem of environmental impact of the petroleum industry remains topical.

Table 3.4

Distribution of oil refinery accidents number
in terms of process equipment categories [ABROSIMOV, 2002, P. 81]

Equipment	Number of accidents, %
Process pipelines	31.2
Pump stations	18.9
Bulk capacity vessels (heat exchangers, dehydrators)	15.0
Furnaces	11.4
Rectification, vacuum and other columns	11.2
Industrial sewage	8.5
Petroleum tank parks	3.8

The causes of combustible materials release, in most cases, are attributed to product leakage through poorly sealed joints of equipment components, as well as equipment corrosion, faulty assembly, operating procedure violations, and others. Table 3.4 illustrates the oil refinery accidents ratio for different types of equipment.

Main hydrocarbon flows from oil refineries into the environment come from the processing area and petroleum tank parks. Atmospheric pollution occurs as a result of process emissions, soil and groundwater pollution is due to pipelines and tank bottom leaks, and surface water pollution is due to the release of untreated effluent water.

In areas of light petroleum products thermal units (gasoline, kerosene, diesel fuel), total hydrocarbons quantity in the upper 20 cm of the soil profiles is relatively low due to their evaporation, oxidation, and migration deeper into the soil. However, large quantities of residual polycyclic aromatic hydrocarbons (including carcinogenic benzo[a]pyrene) are accumulated in the soils of this area.

Soils and sediments in the effluent treatment area are characterized by the elevated general level of hydrocarbon pollution. Hydrocarbon anomalies are distributed across the entire area that surrounds the effluent treatment plant. The anomalies are formed by accumulations of heavy petroleum products with various oxidation degree, and free gases, among which methane is absolutely predominant. Sediments with the petroleum products content of 1,000–35,000 mg/kg are confined to backfilled pits, tanks in the bitumen unit area, effluent treatment plant tanks, and oil traps. The localized phenomenon of intensive gas generation could be attributed to evaporation of the liquid petroleum products that migrate in the capillary fringe zone above the water-bearing horizons.

Geochemical anomalies are formed in soils rocks. They indicate subsurface leaks of petroleum products. In buried pipeline areas, the anomalies are characterized by heavy saturated and unsaturated hydrocarbons being predominant in gases and by presence of relatively light petroleum products and carcinogenic PAHs. Pollution in the petroleum tank park area is determined by methane predominance in the gaseous component and by the ground being heavily polluted by residual fuel oil. Carcinogenic PAHs are not common. Gas presence is related to degassing of residual fuel oil leaking into ground via tank bottoms.

The stream of heavy petroleum products that results from residual fuel oil storage leaks or from spilling this product on the ground surface pollutes with heavy fractions mainly the upper 3-m-deep layer.

In the petroleum tank park area flows of light petroleum products like kerosene and diesel fuel are observed at depths of 4–6 and 7–9 meters migrating above the groundwater table.

Thus, during the years of its continuous operation, the plant creates a huge ground mass saturated by petroleum products underneath. This ground serves as a permanent source of surface and groundwater pollution in the surrounding territory.

In the refinery territory, petroleum products and their production waste penetrate down to the groundwater horizon, move with groundwater, and discharge into the nearest water courses, including major rivers. A water well constructed in the vicinity of a refinery quickly fills up with pure gasoline or diesel fuel, if they are located at shallow depths.

These fuels could be put straight into vehicle tanks. Light fractions of petroleum products and gases often leak in the first terrace above the floodplain. If ignited, they turn into "eternal flames" marking the discharge of polluted groundwater along the river course.

3.5. THE ENVIRONMENTAL IMPACTS OF HYDROCARBONS FROM OIL, GAS, AND PETROLEUM PRODUCTS STORAGE

Crude hydrocarbons and petroleum products storage facilities affect the natural environment one way or another, being a source of oil and petroleum products anthropogenic flows (Fig. 3.1).

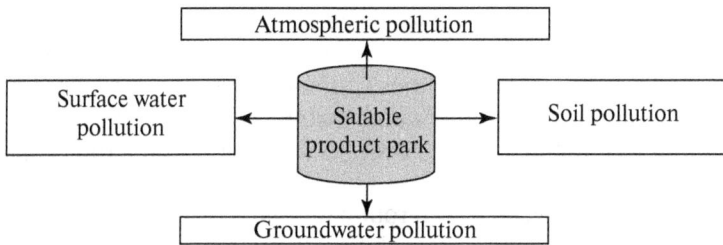

Fig. 3.1. Directions of salable product parks environmental impact.

Petroleum tank parks, which are oil and petroleum products storage facilities located on special industrial sites, are used throughout the supply chain from an oil field to an oil refinery to the end user. They are one of the main sources of anthropogenic petroleum product flows. Firstly, oil and petroleum product leaks from tanks occur due to bottom corrosion or faulty assembly. These leaks are most often hard to observe, but they produce a strong combined effect. If, having reached groundwater or perched water level, hydrocarbons migrate to discharge locations, it creates a hazard of ground and surface water permanent pollution. The oil accumulating in soils and sediments under the tank and releasing light fractions and gases creates a fire and explosion hazard in a tank.

Other process facilities are also located on petroleum tank sites. These are oil, petroleum product and gas pipelines, oil and gas separators, effluent treatment plants, pumps with various functionality, product-metering stations, loading racks, etc. In total, they are responsible for several tons of emissions per year.

Geochemical studies indicate that tank sites are characterized by ground pollution on 75–90% of their area. It should be noted that not only elevated and high concentrations of petroleum products in the ground but also abnormal hydrocarbon concentrations in the soil atmosphere are observed. Besides, natural gases, dangerous hydrogen sulfide concentrations are also present in the soil atmosphere. Elevated petroleum products pollution is also observed in the drain channel water.

Underground petroleum products and natural gas storage

Underground gas storage is a production process that involves gas injection, recovery, and storage in underground tanks constructed in cavities of low-permeability rocks (rock salt, clay, frozen ground). Reservoir beds in natural traps of depleted fields or fields initially filled with water are also used for this purpose.

Gases in free cavities of rocks are stored mostly in a liquefied state at room temperature and a pressure of about 0.8–1 MPa (8–10 kgf/cm^2) and more. Normally, these are propane, butane, or a mixture of them. Underground gas storage facilities can hold hundreds of millions cubic meters (sometimes, billions of cubic meters) of gas. More than 600 underground gas storage facilities operate worldwide with a total capacity of about 340 billion m^3.

Main disadvantage of underground storage is that it is not tight. A risk of petroleum products loss through the tank walls and substance dispersion in subsurface water and sediments is always present. Underground storage reliability largely depends on the geological state of the rocks where this storage facility is constructed, and this factor is not always under control.

Salt is viewed as the most reliable screen for liquid and gas. Storage facilities in the salt are constructed at a depth of 80–100 to 1,000 meters and more. In terms of environmental safety, construction of salt storage facilities is a relatively cheap and the least environmentally damaging way of storing petroleum products. On the other hand, such storage facilities are extremely short-lived. Rocks pressure gradually makes the salt cavity fill up, while petroleum products and residual salt-water solutions are forced up to the surface and pollute the environment.

However, underground storage facilities are much safer in terms of fire and explosion hazard than aboveground facilities. Besides, the storage area does not take up land territory, and consumption of large metal quantities for building the tanks is avoided.

CHAPTER 4
ATMOSPHERIC HYDROCARBON POLLUTION

4.1. CHANGES OF THE ATMOSPHERIC BOUNDARY LAYER
ON OIL AND GAS FIELDS AND IN THE VICINITY
OF OIL AND GAS PIPELINES

Hydrocarbon flows to the atmosphere can originate from numerous sources. They include oil and gas pipelines, exploration and production wells, compressors, diesel engines, pump stations, salable product parks, oil, gas, and condensate field treatment units. Gaseous hydrocarbon leakage from these sources seriously affects the state of the atmosphere.

More than 80% of all hydrocarbon emissions to the surface atmospheric layer come from main gas pipelines, primarily, due to their length. Hydrocarbons are released during scheduled pipeline repairs, connecting new sections, and because of leaks in gas compressor units. In the north of Western Siberia, methane emissions intensity in the gas field regions amounts to 2.1 ± 0.3 million tons per year. According to calculations, this is equivalent to 380 mg/m^2 per day, while emissions of natural sources in this area are 10–20 mg/m^2 per day. The amount of methane released into the atmosphere from wells and pipelines is 67% to 100% of total methane emissions to the atmosphere with the background methane concentration of 1.82 vol. % [RESHETNIKOV ET AL., 2006].

Flare units burning associated petroleum gas make up numerous sources of atmospheric pollution in the fields. This problem is especially relevant in the fields where a developed infrastructure of associated gas transportation and disposal does not exist. Besides flare unites, incomplete hydrocarbon combustion products are released from pits through exhaust and smoke stacks. Hydrocarbons are released to the atmosphere during blowing of wells, pipelines bleeding-off, leaks from untight process units, evaporation from effluent treatment plants and salable product tank parks.

The main products produced by natural gas combustion are carbon oxides (CO_2 and CO), nitrogen and sulfur oxides (NO_2, SO_2), and benzo[a]pyrene. If combustion ends spontaneously, methane and heavy hydrocarbons are discharged into the atmosphere, and later condense in the atmospheric boundary layer and precipitate on the soil and vegetation.

Among carbon oxides, the colorless and odorless carbon monoxide (CO) is the most dangerous. It is produced during gas combustion under oxygen deficit conditions. If its concentration in the air mixture is higher than 12.5%, carbon monoxide ignites. Nitrogen oxide is formed from molecular nitrogen present in the atmosphere or natural gas, by high-temperature combustion. Sulfur oxides are released if hydrogen sulfide is present in gas or the atmosphere. Formation of benzo[a]pyrene and other PAHs occurs due to incomplete hydrocarbon combustion, especially, if heavy hydrocarbons or condensate is present in gas, besides methane. In this case, a thick cloud of soot appears where benzo[a]pyrene is concentra-ted. Information about oil and gas field impact on the atmospheric boundary layer, as illustrated by the Khanty-Mansiysk Autonomous Region in Western Siberia [MOSKOVCHENKO, 2013]. According to the author's information, air below flare units contains twice the amount of nitrogen dioxide and soot as compared to conventionally background level areas, and 1.3 times more carbon oxide and methane. In the areas of exploration and production wells impact, the amount of soot is 2 times higher compared to the background level, the amount of nitrogen dioxide is 1.5 times higher, and sulfur and nitrogen dioxide — 1.3 times higher. Atmospheric fallout observed in the snow cover is a reliable indicator of atmospheric conditions in the winter period. With the equal quantity of aerosols, their composition in the fields is substantially different from background levels: nitrate nitrogen quantity is 3 times higher, nickel and chrome quantity is 4–5 times higher, and that of zinc and mercury is 2–4 times higher. Even more profound changes of the surface atmospheric layer take place in residential areas. The suspended solids quantity increased there by 26 times. Compared to background level areas, phenol content in suspended solids is 3 times higher, ammonia nitrogen — 5 times, mercury — 7 times, iron — 17 times, petroleum products — more than 20 times higher. Aerosols qualitative composition in the fields and in residential areas differs in terms of nitrate and ammonia nitrogen, phenol, iron, and petroleum products content. In the fields, aerosols have higher content of heavy metals (nickel, chrome, copper), and lower content of mercury.

4.2. IMPACT OF OIL REFINERIES
ON THE GROUND-LEVEL ATMOSPHERE

Oil refineries release more than 1.5 million tons of harmful substances per year — hydrocarbons (about 80%), sulfur oxides, hydrogen sulfide, carbon oxides, nitrogen oxides, phenol, and solids to the atmosphere. Because emission of substance into the atmosphere is inherent to the production process, refineries are constructed on high, well-aerated areas. In this case, harmful substances are dispersed across a vast space and subsequently precipitate onto soil and water body surface.

The major sources of atmospheric pollution in a refinery are:

- tanks for storing oil, petroleum products, and various toxic low boiling liquids;
- effluent treatment plants; certain process units (AVT, catalytic cracking, bitumen production, etc.);
- flare systems.

The highest atmospheric hydrocarbon concentrations are observed in the areas where these facilities are located.

In the vicinity of large oil refineries, consistently high content of atmospheric pollutants is observed. It falls very slowly with the growing distance from emission source. Emergencies result in the most dangerous situations.

Environmental pollution degree depends on the volume of processed oil, the employed technology, and equipment condition.

Hydrocarbons and sulfurous gas account for more than 85% of oil refinery emissions to the atmosphere. The largest volume of hydrocarbons is released to the atmosphere by the refinery oil and petroleum products storage tanks. Hydrocarbons are relieved through special valves in case of overpressure (in- and out-breathing), and via manholes. Under tanks normal operating conditions, 80% of hydrocarbons enter the atmosphere during "out-breathing" (at the time of tank filling or emptying), while 20% is released during "in-breathing" (daily operation of pressure control valves).

4.3. OTHER SOURCES OF THE ATMOSPHERIC BOUNDARY LAYER POLLUTION

Localized areas of the hydrocarbon pollution of the atmospheric boundary layer that are not related to petroleum industry are observed in large cities with heavy traffic, heat and power stations, landfills, and industrial facilities. The atmosphere and atmospheric fallout in cities contain all classes of hydrocarbons (methane, ethylene, aliphatic hydrocarbons, benzene, toluene, polycyclic aromatic hydrocarbons, phenol, formaldehyde, etc.).

By the beginning of the 21st century, 25 million tons of gasoline and 40 million tons of diesel fuel were burned in Russia annually. As a result, 1.4 million tons of carbon monoxide, 1.8 million tons of nitrogen oxides and 1.6 million tons of hydrocarbons were released to the atmosphere [DANILOV, 2003].

Atmospheric pollution by polycyclic aromatic hydrocarbons (PAHs) is particularly important, as they include compounds with carcinogenic and mutagenic properties. The formation of PAHs in anthropogenic processes is related to pyrolysis of carbonaceous feedstocks (coal, oil, wood, etc.) under high temperatures and oxygen deficit. The temperature of PAH mass formation is 650–900 °C, while PAH can be also formed under lower temperatures (300–350 °C).

Main sources of PAHs emission to the atmosphere are heat and power stations, coke production facilities, and facilities heated with coal. These sources account for more than 85% of anthropogenic PAH total weight released to the atmosphere. Exhaust gases of road vehicles and other machines and units contribute a lot to pollution of the atmosphere by PAHs. In the hydrocarbon composition of exhaust gases, PAHs occur along with paraffins, olefins, and mononuclear aromatic hydrocarbons.

Benzo[a]pyrene is the most common carcinogenic hydrocarbon among PAHs. Its maximum allowable concentration in the atmosphere is $1.5 \cdot 10^{-5}$ ng/m^3. Benzo[a]pyrene is viewed as a general carcinogenic activity indicator.

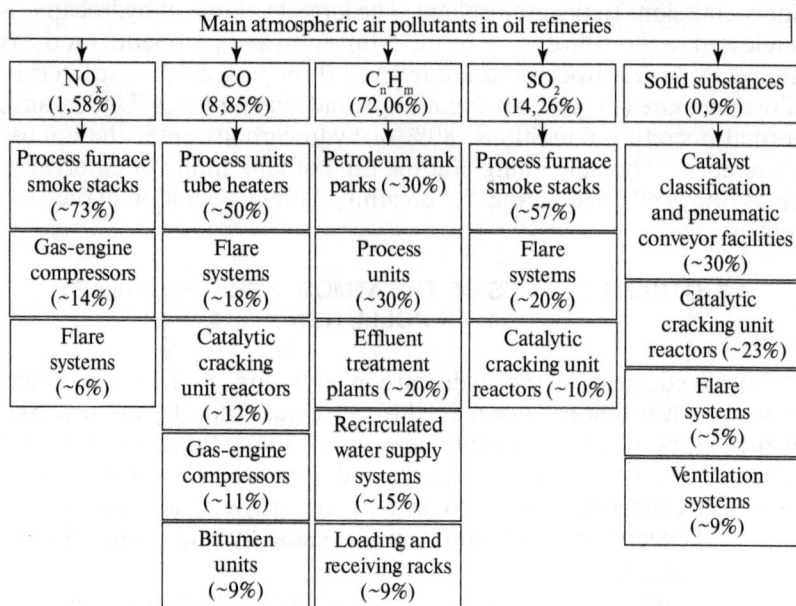

Main atmospheric air pollutants in oil refineries				
NO$_x$ (1,58%)	CO (8,85%)	C$_n$H$_m$ (72,06%)	SO$_2$ (14,26%)	Solid substances (0,9%)
Process furnace smoke stacks (~73%)	Process units tube heaters (~50%)	Petroleum tank parks (~30%)	Process furnace smoke stacks (~57%)	Catalyst classification and pneumatic conveyor facilities (~30%)
Gas-engine compressors (~14%)	Flare systems (~18%)	Process units (~30%)	Flare systems (~20%)	Catalytic cracking unit reactors (~23%)
Flare systems (~6%)	Catalytic cracking unit reactors (~12%)	Effluent treatment plants (~20%)	Catalytic cracking unit reactors (~10%)	Flare systems (~5%)
	Gas-engine compressors (~11%)	Recirculated water supply systems (~15%)		Ventilation systems (~9%)
	Bitumen units (~9%)	Loading and receiving racks (~9%)		

Fig. 4.1. Hazardous substance concentrations in oil refinery gas emissions [Kapustin, Gureyev, 2007, p. 318].

CHAPTER 5
ENVIRONMENTAL IMPACTS OF OIL PRODUCTION
ON SOIL AND VEGETATION COVER

5.1. MECHANICAL IMPACT ON SOILS AND VEGETATION

All phases of oil production, i.e., oil exploration, field construction, production, transportation, and primary processing, result in soil and ground mechanical damage and produce a variety of secondary anthropogenic landscape phenomena (Fig. 5.1).

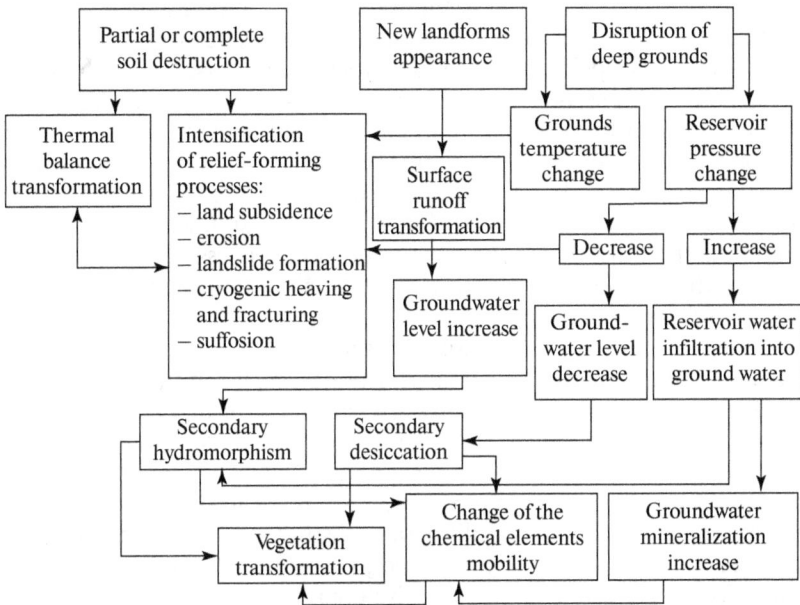

Fig. 5.1. Consequences of soil mechanical disruption
in oil production areas.

Mechanical damage results in anthropogenic landforms, both positive (mounds, embankments, dumps), and negative (pits, trenches, quarries, etc.). Terrain transformation is associated with runoff conditions modification, flooding or desiccation of territories, increase in erosion activity, formation of gullies, subsidence, landslides, and ravines.

Mechanical impacts disrupt soils integrity or bury them under a layer of anthropogenic deposits. Resulting landscape changes can be both reversible and irreversible, depending on the intensity of soil and ground mechanical disruption and the severity of soil and geochemical processes modification. The number and hazard level of secondary processes, mechanical in origin, could significantly outweigh primary landscape changes, while the area of damaged territory could be much larger than the land allotment area.

The specific features of relief and relief-forming process change caused by mechanical soil disruption vary depending on landscape conditions.

In cryogenic tundra-taiga landscapes characterized by a wide presence of permafrost, all forms of cryogenesis actively develop: thermokarst, thermal erosion, solifluction, frost heaving, frost cracking, and increased soil thixotropy and liquefaction. These processes occur in arctic tundra, tundra, forest-tundra, northern and middle taiga landscapes, as well as in certain southern taiga landscapes with permafrost. For some landscapes in this group, cryogenic processes are manifested by the activity of stone rivers and screes. Erosion in the broadest range of its forms can be initiated in all climatic zones, from southern taiga plateaus with isolated permafrost to subtropics, including mountainous areas. Imost landscapes, disruption of soil and vegetation cover creates a risk of sheet and ravine erosion. Catastrophic erosion with badland formation is the most dangerous type of erosion process. These processes are most hazardous in semideserts, deserts, and mountainous grassland, forest, and dry steppe landscapes.

Intensification of wind erosion (deflation) with blowing out surface soil material and creation of eolian landforms occurs, when the soil and vegetation cover integrity is disrupted in steppe and semi-desert landscapes.

Vegetation cover damage or destruction by heavy vehicles is particularly intense during oil/gas field construction. They result in appearance of anthropogenic meso-, micro- and nanorelief forms, intensification of water and wind erosion, solifluction, disturbance of surface and subsurface, and, in cryogenic landscapes, change of the soil temperature conditions. One runoff of the negative consequences of the mechanical disturbance of soils in oil production areas consists in formation of the extensive sand blowouts characterized by very slow revegetation, especially in the North.

A distinctive feature of the mechanical soil disturbance in humid landscapes is the development of secondary hydromorphism and expansion of swampy areas. If secondary swamping coincides with the natural direction of landscape development, as, for example, is the case in Western Siberia, then it is irreversible.

5.2. OIL BEHAVIOR IN SOIL

Radial (vertical) oil migration in the soil profile

Oil enters the soil from many sources. The heaviest pollution occurs from uncontrollable flowing of wells and pipeline ruptures. Oil from wells is released onto the soil surface as part of the produced fluid that includes associated reservoir salty water and brines. The environmental consequences of soil pollution depend on the quantity of pollutant substance, duration of pollution process, produced fluid composition, the nature of oil and salts migration and distribution in the soil profile, and soil tolerance towards oil pollution and salinization.

Oil movement in the soil profile depends on the following factors: (1) oil properties (its density and viscosity), duration and type of its inflow (surface or subsurface), (2) soil properties, and (3) landscape-geochemical position of the soil.

The crucial properties of the soil that determine oil radial travel characteristics are oil capacity, the system of subsurface geochemical barriers and migration channels, and the texture and water regime of the soil. Oil capacity of a soil is the maximum possible content of oil in the soil mass for each moisture level [SOLNTSEVA, 1998, P. 69].

High oil capacity is characteristic of organic (peat, muck) and organomineral (humus) soil horizons. Mineral substrates of the coarse texture (sand, sandy loam, gravelly deposits) have the lowest oil capacity, therefore they are penetrated by oil to the greatest depths.

Geochemical barriers in soils

Oil accumulation levels and its distribution in the soil profile are also greatly affected by subsurface geochemical barriers. According to A.I. Perelman, geochemical barriers are "Earth's crust areas where a sharp decrease in the intensity of migration of chemical elements occurs within a short distance, resulting in the decrease of their concentration" [PERELMAN, KASIMOV, 2,000, P. 45]. Soil profile is essentially a system of geochemical barriers. The following act as radial barriers and oil accumulators: surface litter (O), sod (upper soil horizon densely penetrated by roots), muck (highly decomposed organic material, H), peat (O, T), humus (A), and illuvial (B) soil horizons. Oil concentrations in the litter and humus horizons can reach 10–20% of the soil weight in case of surface inflow of oil. Particularly high concentrations of bituminous substances are typical for peat horizons (see Fig. 5.2C).

Some soil horizons are essentially screen barriers with a very low permeability for oil due to extremely small-sized pores and capillaries. They include cryogenic, gley, clayey illuvial, and illuvial-gleyic horizons. Oil hydrocarbons are accumulated above these horizons (see Fig. 5.2). Howe-ver, these barriers are not absolutely impermeable. Their imper-meability is disrupted by cracks, wormholes, root channels and, also, depends on duration of contact with the pollutant.

The groundwater horizon is a kind of a barrier where oil transits into the lateral soil and groundwater runoff. If the source of oil is a subsurface one, a strongly pronounced maximum of oil hydrocarbons concentra-tion is formed above the groundwater table.

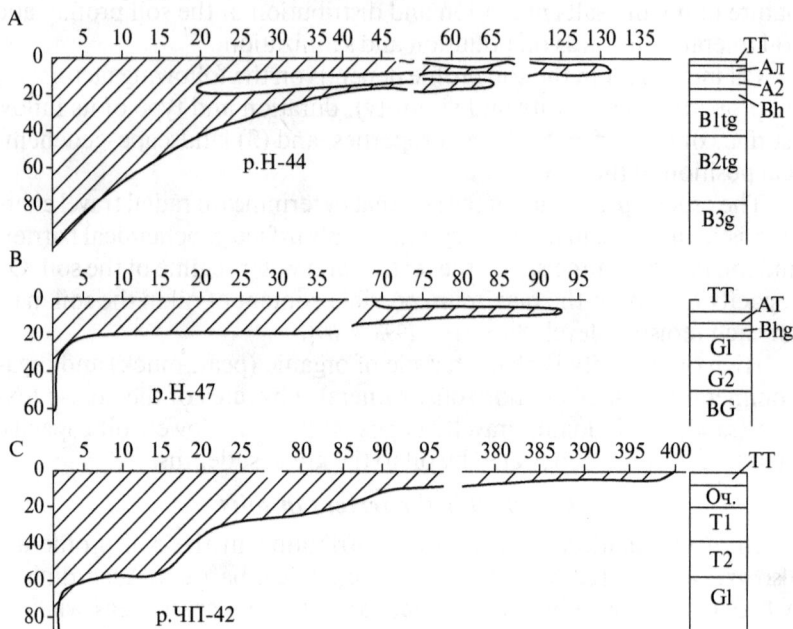

Fig. 5.2. Possible radial oil distribution (g/kg)
in the vertical soil profile after pollution in:
A — iron-illuvial podzols (Rustic Albic Podzols); B — tundra gley soils
(Oxyaquic Cryosols); C — shallow peatlands (Histosols)
[SOLNTSEVA, 1998, P. 75].

If several oil barrier-accumulators and barrier-screens are present in the soil profile, several maxima of oil hydrocarbons can be observed. This is, for example, typical of oil-polluted Podzols (see Fig. 5.2A) where one of the maxima is confined to the humus-accumulative horizon with the highest oil capacity, while the second maximum is formed over the barrier-screen, i.e. the low-permeability iron-illuvial (often, with stagnic features) horizon.

Landscape-geochemical position of the soil. Radial oil migration is predominant on interfluves, within autonomous landscape-geochemical positions. Their maximum of oil hydrocarbons is typically in the upper soil horizons.

The importance of lateral substance migration increases in the soils of slopes in transitional landscape-geochemical positions. On steep slopes in the presence of actively functioning migration channels, lateral oil runoff into lower-lying geochemically conjugated relief positions prevails. Soils of superaquatic accumulative landscapes (floodplains) receive oil hydrocarbons coming from the soils of higher hypsometric layers, as well as from the groundwater. With the general direction of substance runoff towards the local accumulation area, these soils often act as lateral geochemical barriers. Superaquatic position is often occupied by bog soils that are characterized by a very high level of oil pollution of their organic horizons despite their high water content.

Lateral oil migration in soils and the boundaries of oil pollution

Lateral oil migration from pollution sources creates extensive areas of local pollutant accumulation "contamination aureoles" in soils, aquatic environment, and bottom sediments. N.P. Solntseva [1998] developed the concept of lateral migration of oil and salts in the soils that are referred to as "oil and salt contamination aureoles."

Formation of contamination aureole. When oil travels through the soils laterally, it is fractionated. This results in formation of areas with different levels of pollutant content and composition with a distinctive zonal pattern. Thus, there is the pollution "nucleus" where the bulk of low-mobility high-molecular-weight oil components is concentrated in the vicinity of oil discharge source. In the peripheral parts of contaminated zone (aureole), low-molecular-weight hydrocarbons prevail, while the content of bituminous substances is smaller (see Fig. 5.3).

Pollution boundaries change over time. Over time, pollution aureoles go through the secondary redistribution of bituminous substances and change of geochemical structure. Depending on terrain slopes, surface wash off and subsoil migration of bituminous substances result in their maximum concentrations being shifted to the lower soil horizons and towards the periphery of the original pollution aureole. "Inverted" aureoles are created, where the maximum content of bituminous substances is located in the peripheral part. The highest contrast of "inverted" aureoles is observed, when bituminous substances migrate from mineral soils into the lower-lying geochemically conjugated organomineral and organic soils (see Fig. 5.4).

Regardless of natural conditions, the boundaries of contaminated zone expand over time due to lateral migration of bituminous substances and pollution of soil and groundwater.

5.3. ANTHROPOGENIC SOIL SALINIZATION

Soil salinization, as well as its pollution by oil, is typical for petroleum-producing areas. In the oil field area, the salts come from crude oil, field wastewater, pits content, flushing liquids, etc. Salt composition is normally dominated by chlorides, less often by sulfates and carbonates of sodium, calcium, and (in case of boggy landscapes) aluminum and iron.

Under all natural conditions, very high variability of soil salinization is observed even at short distances. This is caused by landscape heterogeneity, variation of salts quantity and composition during the field operation, and by repetitive salinization incidents.

Salinization intensity depends of the anthropogenic flow volume and mineralization, as well as on the soil properties (horizonation of the profile, texture, etc.).

Formation of anthropogenic salinization aureoles

Radial and lateral salt migration results in formation of primary aureole of salt contamination; the fractionation of salts depending on their mobility takes place within zone. The highest salt content is confined to the salinization "nucleus" (see Fig. 5.5), where, even under humid climatic conditions, anthropogenic solonchaks (> 1% of soluble salts) or saline soil varieties (with 0.25−1% of soluble salts) are often formed.

Near the source of anthropogenic saline flow, the least mobile carbonates, iron, and bituminous components are accumulated, and ferrous and bituminous saline soils are formed. Sulfates migrate farther, while the most mobile chlorides intensively move to the lower part of the profile and to the peripheral areas of the pollution aureole. Therefore, regardless of natural conditions, mainly the chloride-sodium salinization is observed in the soils of primary aureole periphery. Large volumes of saline water released into soils result in intensification of reduction processes and formation of waterlogged saline soils. Soil salinization leads to degradation and death of native vegetation.

Anthropogenic salinization boundaries change over time. Under any natural-climatic conditions, the boundaries of saline pollution aureoles are transformed over time. In a humid climate, soils are desalinized. The rate of desalinization depends mainly on the soil water regime and the

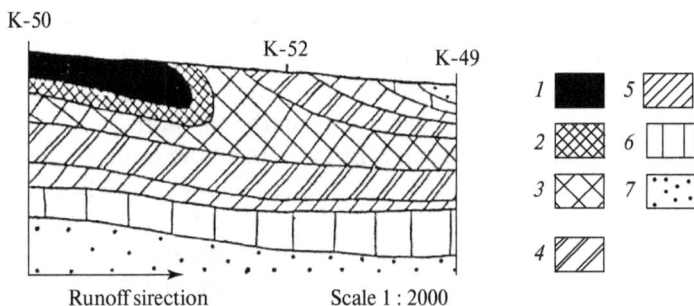

K-50

K-52   K-49

1 ■	5 ▨
2 ▨	6 ⬜
3 ▨	7 ⬚
4 ▨	

Runoff sirection   Scale 1 : 2000

Fig. 5.3. Bituminous substances content (g/kg soil) in tundra gley soils immediately after contamination with crude condensate (forest-tundra, Western Siberia). Bituminous substances, g/kg: *1* — 70–50; *2* — 50–30; *3* — 30–20; *4* — 20–10; *5* — 10–5; *6* — 5–1; *7* — < 1. [SOLNTSEVA, 2009, p. 50].

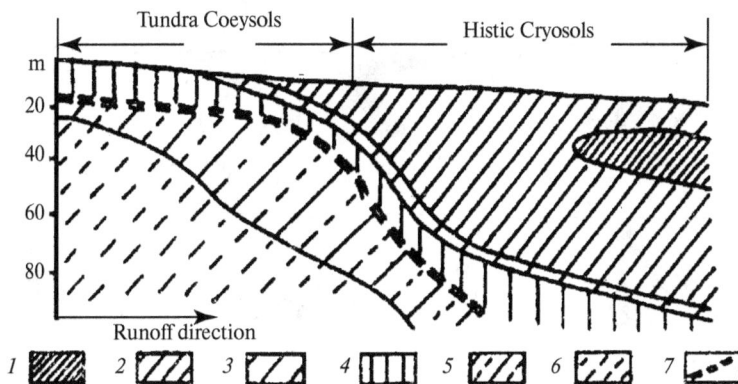

Tundra Coeysols   Histic Cryosols

m
20
40
60
80

Runoff direction

1 ▨ *2* ▨ *3* ▨ *4* ⬜ *5* ▨ *6* ▨ *7* ⬚

Fig. 5.4. Petroleum product content in soils contaminated with untreated effluent water (forest-tundra, Western Siberia) Bituminous substances, g/kg soil: *1* — more than 20; *2* — 20–15; *3* –15–10; *4* –10–5; *5* — 5–1; *6* — < 1; *7* — physicochemical gley barrier [SOLNTSEVA, 2009, p. 52].

composition of anthropogenic salt flow. Under otherwise equal conditions, the highest desalinization rate will be ensured in soils with the leaching water regime. Thus, in soddy-podzolic soils (Albic Retisols), a year after salinization by chloride-sodium water, salt concentrations in the upper horizons decrease by 7–8 times. However, such soils are still categorized as salinized (salts content is 0.25%). Even in 15–18 years, the content of soluble sodium remains 5–6 times higher than the background values. In leached chernozems (Greyzemic Chernozems) with the periodically percolative type of water regime, the desalinization process is going much slower for the same salinization type: a year after salinization the content of soluble salts in the top horizon is up to 1.5%.

To a large extent, the desalinization rate of the same type of soils is determined by the composition of anthropogenic saline flows. For example, bituminous soils contaminated by effluent chloride-sulfate water containing petroleum products are desalinized quite slowly due to a lower mobility of sulfates as compared to chlorides and due to salts binding by oil emulsion. It took 4 years after salinization for the soluble salt content in the top horizons of these soils to decrease by 4–6 times. The desalinization rate of soils contaminated by crude oil is even lower (see Fig. 5.5). Even 20 years since salinization, the content of soluble in the lower horizons of such soils is 7–8 times higher than the background values.

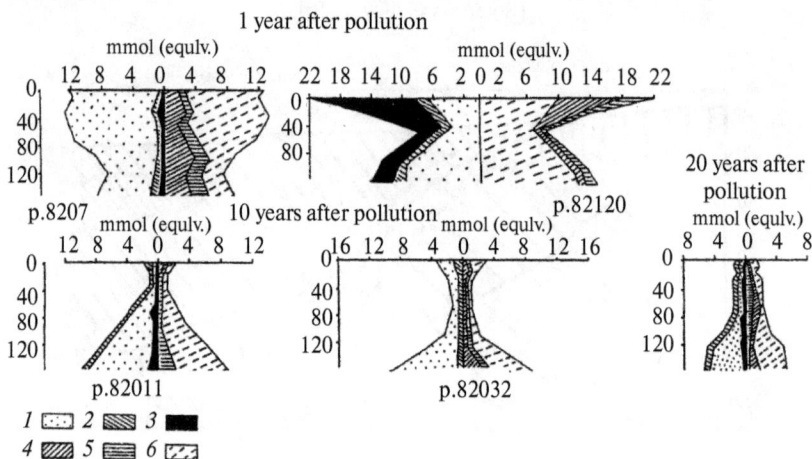

Fig. 5.5. Distribution of soluble salts in the soddy-podzolic forest soils (Albic Retisols) after different periods from the moment of their contamination by crude oil (southern taiga, the Kama region near Perm) [SOLNTSEVA, 2009, P. 70]:

$1 - SO_4^{2-}$; $2 - Cl^-$; $3 - HCO_3^-$; $4 - Ca^{2+}$; Mg^{2+}; Na^+.

Under otherwise equal conditions, desalinization intensity is higher in sandy and loamy sandy soils, as well as in the soils of the arable land. Desalinization processes are strongly constrained in the subordinate landscapes that receive anthropogenic flows from all pollution sources in the catchment.

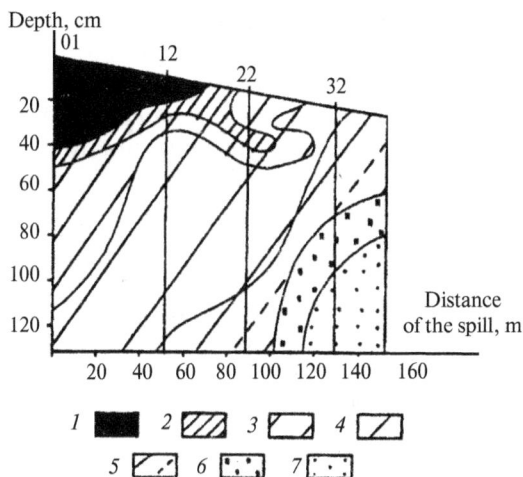

Fig. 5.6. Soluble salts distribution in soddy-podzolic soils (Albic Retisols) contaminated by effluent water; two weeks after contamination (southern taiga, the Kama region near Perm). Total of salts (%): *1* — > 1.0; *2* — 1.0–0.75; *3* — 0.75–0.50; *4* — 0.50–0.25; *5* — 0.25–0.10; *6* — 0.10–0.05; *7* — < 0.05 [SOLNTSEVA, 2009, P. 67].

Soil desalinization involves shifting of the maximum salt content from the top to lower soil horizons, from the "nucleus" to the periphery of contaminated zone, with the maximum salt content going beyond the limits of the primary salinization area leading to the formation of solonchaks and saline soil varieties (see Fig. 5.6). Secondary aureoles of soil contamination with soluble salts are formed.

Sodium chlorides and ions that predominate in the lower soil horizons and in the outer "lining" of secondary saline aureoles migrate with the highest intensity. Sulfates are recorded in the middle part of the soil profile and in the middle part of secondary salt aureoles. The least mobile calcium bicarbonates are typically found near the source of anthropogenic flow discharge.

Multiple discharge of mineralized water onto the same territory results in formation of heterogeneous salinization aureoles with complex spatial distribution patterns of individual ions.

Therefore, one can identify the following distinctive features in the formation of anthropogenic oil and salt contamination aureoles.

1. Regardless of pollutants composition, the boundaries of anthropogenic contamination aureoles change over time.

2. When the substances of anthropogenic flows interacts with the soil material, the former are fractionated creating a distinctive geochemical zoning within the contamination limits.

3. The level of pollutants content and geochemical zoning structure of contamination change over time forming "inverted aureoles", while the primary contamination outline expands.

5.4. SOIL PROPERTIES CHANGE DUE TO THE IMPACT OF PRODUCED FLUID

The produced fluid actively engages with the soil material and leads to transformation of physical and chemical properties of the soil exposed to anthropogenic flows. Below, we review the change in physical, chemical, and physicochemical properties of soils in areas of oil production and transportation.

Hydrocarbon state of the soils

Oil pollution changes the hydrocarbon state of the soils, i.e. the quantity and composition of natural gases in the soil atmosphere, as well as of bitumoids and individual hydrocarbon compounds (polycyclic aromatic hydrocarbons and normal alkanes). Unpolluted zonal soils typically have low content of all soil hydrocarbons. Normally, methane predominates in their natural gases composition and odd homologues C_{25}–C_{33} predominate in their normal alkanes composition. They are characterized by light hydrocarbon type of bitumoids and phenanthrene-naphthalene association of polycyclic aromatic hydrocarbons [PIKOVSKY ET AL., 2008; GENNADIEV ET AL., 2015B].

Surface oil and petroleum product spills result in the injection hydrocarbon state of the soils that differs from the background hydrocarbon state by a very high content of all hydrocarbon groups. Heavy hydrocarbons and sulfurous gases appear in the gaseous phase, while the bitumoids composition becomes similar to that of crude oil. Their soil content increases by hundreds of times. PAH levels in oil-polluted soils exceed the background values by more than 200 times. Among them, tetraphene, benzo[ghi] perylene, chrysene, pyrene, and perylene are found. In the composition of normal alkanes, the fraction of compounds with an even number of carbon atoms dramatically increases, while the even and odd alkanes ratio becomes close to one.

Humus state of the soils

Oil and petroleum products that have been released into the soil interact with its organic and mineral compounds. This results in the change of both total organic carbon content and the composition of the soil organic matter and its distribution in the soil profile.

After oil is released on the soil surface in any natural zone, the soil organic carbon (C_{org}) content is dramatically increased, especially in the upper horizons. Moreover, the highest contrast of C_{org} distribution in the profile is typical of peat and peat-gley soils, where the gley horizons maintain very low values of C_{org}. In the taiga zone, at an equal distance from pollution source, deeper penetration and higher content of oil is observed in the soils of divide surfaces and mild slopes (as compared to the soils of steep slopes). However, the maximum levels of oil accumulation have been recorded in the soils of hollows.

Organic carbon spatial distribution depends on the landscape variety, type of land use, anthropogenic landforms in the areas of oil contamination, and the repetition of contamination incidents.

The composition of organic matter in contaminated soils is changed. The carbon-to-nitrogen ratio substantially increases, as well as the content of nonhydrolyzable residue. As demonstrated by simulation field experiments, the fraction of humic acids in the composition of humus decreases within 2–5 years after oil contamination in mineral soils with the oxidizing conditions [Oborin et al., 2008].

Water and air regime and oxidation-reduction conditions in the soil profile

Disturbance of the soil water and air regime is one of the reasons why oil contamination negatively affects terrestrial ecosystems. It is mainly related to oil filling the free pore space in soils. If the concentration of viscous resinous components in anthropogenic flows is high, over time the soil surface becomes covered with a bituminous crust. Oil penetration into the soil results in giving it hydrophobic properties. The oil hydrophobic molecules envelop plants roots, leaves, and stems and impair the process of air and water exchange between the plants and the environment.

In all natural zones, reduction processes intensify (or appear) in oil-contaminated soils, to the extent of creating a hydrogen-sulfide setting. These processes are the most active in the polluted soils of tundra and forest-tundra landscapes.

Reduction conditions also appear when additional volumes of liquid from anthropogenic flows are released into the soil.

The secondary intensification of reduction conditions occurs in a few months after contamination: gray-blue, blue, blue-green, and olive gley horizons appear in the soil profile. Reduction process intensity grows over time. Often, this leads to the vegetation being replaced by hydrophilic and even bog vegetation. The portion of gley soils in the soil cover increases.

Soil exchange complex (SEC)

As sodium salts normally predominate among the soluble salts of oil field anthropogenic flows, it is typical for the Na^+ ion to be implanted in the SEC of polluted soils. This leads to the change in the composition of adsorbed cations. Such changes are the most dramatic in the soils with the low base saturation (such as some tundra-gley (Oxyaquic Cryosols), podzols (Albic Podzols), and podzolic (Retisols) soils). The fraction of exchangeable Na+ in contaminated soils may amount to 25–40% and more of the total adsorbed cations (solonetzic process) resulting in soil alkalization. In the case of base-saturated soils (Chernozems, Kastanozems, etc.) secondary formation of gypsum and carbonates takes place.

With the recovery of soil microbiota and intensification of oil degradation and decomposition of plant residues, the content of H^+ ion in soil solutions increases. Soil desalinization begins: secondary adsorption of hydrogen ions by the SEC occurs. This results in restoration of the SEC composition in the unsaturated soils and further SEC transformation in the initially base-saturated soils. As a result of exchange reactions with SEC cations, the soil solution receives predominantly Na+ ions with the formation of sodium bicarbonate. This leads to strong alkalization of the soil.

Desalinization processes are most intense in the upper soil horizons of the pollution nucleus. At the same time, an intensive alkalization takes place in the lower soil horizons and in the margins of contamination zones. Eventually, the maximum manifestation of soil alkalization is moved beyond the limits of original contamination area. A period of 20–25 years is not sufficient for SEC recovery in initially unsaturated soils of the southern taiga subzone that have been transformed by anthropogenic factors. For initially base saturated soils, SEC transformation could be irreversible.

Acid-base soil conditions

Soil anthropogenic alkalization and dealkalization processes result in changes of acid-base conditions. At the initial stage of soil transformation, alkalization if the upper horizons occurs in the nucleus of primary contamination aureole. In the middle and marginal zones of contamination aureoles in the taiga zone, acidification could be observed in the entire soil profile. Eventually, under any natural conditions, within the limits of initial contamination aureole the soil is alkalized across the entire profile. The highest alkalinity is confined to the aureole nucleus.

The most stable acid-base conditions exist in the soils with the base-saturated adsorption complex. In base-unsaturated soils, organic horizons are relatively more resistant to pH changes, while mineral horizons are less resistant. The intensity of changes in the acid-base conditions is higher for soil contamination with oil than for contamination with wastewater.

Further stages of anthropogenic soil transformation involve the decrease in the alkalinity of upper soil horizons in the contamination nucleus because of the loss of sodium from the adsorption complex (dealkalization). The maximum of alkalinity is shifted to the lower horizons and to the contamination area periphery.

Acid-base conditions in the oil-contaminated soils of the taiga zone remain modified for dozens of years. Repeated anthropogenic flows complicate the acid-base zoning within the contamination aureoles.

Migration capacity of elements and bulk elemental composition of soils

The change in acid-base and oxidation-reduction conditions in contaminated soils of oil production areas results in changes of migration capacities of macro- and microelements. According to N.P. Solntseva, the combination of acid and slightly acid soil reaction with secondary soil gleyzation in contamination aureoles enhances the mobility of many elements (Fe, Mn, Al, Cu, Mo, Zn) in the taiga zone. The content of their mobile forms increases in contaminated soils. This is also caused by the additional input of these elements with anthropogenic flows.

Subsequent soil transformation stages that are characterized by the combination of gleying and alkalization result in decreased mobility of certain elements, including iron and aluminum. As a result, they are fixed in soils. At the same time, under any natural conditions the content of silicon and manganese mobile forms in soils increases substantially, and their intense radial and lateral migration is observed. The most intense migration of silicon occurs in oil-contaminated soils. Absolute loss of this element from the eluvial (albic) horizons of oil-contaminated soddy-podzolic soils (Albic Retisols) in the central part of contamination aureole may exceed 6% over a period of 10 years.

The change in migration capacity of macroelements in contaminated soils leads to changes in their bulk content. It is typical for the bulk content of Si and Mn to decrease, and for the bulk content of Al to increase in the soils subjected to the anthropogenic salinization. Under otherwise equal conditions, crude oil contamination results in the most pronounced changes of the bulk elemental composition of the soils.

Soil physical properties

Regardless of natural and climatic conditions, intensive soil pollution results in dramatic changes of its physical properties.

In soils that are exposed to anthropogenic salinization, an increase in the degree of dispersion of the soil mass and the formation of coarse lumps after the soil drying are observed. Such soils become very dense in dry condition. In contaminated soddy-podzolic soils (Albic Retisols), bulk density of topsoil exceeds that in the uncontaminated soils by 30%. The compaction of contaminated soils is observed within the primary and secondary contamination aureoles. In the latter case, the changes in the soil SEC and acid-base conditions over time are important. A tendency for the development of columnar structure has been identified. The intensity of these processes depends on the anthropogenic flows composition, the time elapsed since the moment of contamination, position relative to the contamination source, soil type, and the composition of parent material. The changes of physical properties are the most pronounced in contaminated soils of heavy texture.

5.5. CHANGES OF SOIL BIOCENOSES UNDER THE IMPACT OF OIL AND MINERALIZED WASTEWATER

Oil contamination of soil and dramatic modification of its physico-chemical properties results in rapid transformation of the composition of all soil biota components. Soil dwellers include representatives of different systemic groups of organisms: plants (soil algae), animals, fungi and yeasts, actinomycetes, and bacteria. In the soil, they constitute a single system that ensures normal functioning of the soil and execution of its biogeocenotical functions.

Soil microorganisms

The effect of oil on microorganism communities depends on its content in the soil. Four levels of soil contamination are identified in the gradient of pollutant concentrations. A unique set of changes in microbial communities corresponds to each level [GUZEV ET AL., 1989]. A low-level oil contamination of soil does not result in changes of microorganism species composition. Quite often, hydrocarbons produce a stimulating effect on microorganisms, leading to a slight increase of their total biomass. The range of pollutant concentrations that corresponds to these changes is called the *homeostasis zone*. Higher pollutant concentrations (*stress zone*) result in the changes of species dominance degree in microbial communities: the species, whose quantities in the original soil were small, begin to dominate. At the same time, the soil water and air regime is disturbed. High pollutant concentrations (*resistance zone*) lead to changes in microbial community species composition: death of sensitive species and predominant development of resistant species. Oil becomes

80

the main trophic substrate for hydrocarbon-oxidizing microorganisms (HOM). The latter include the following bacteria genera: *Arthrobacter, Bacillus, Brevibacterium, Nocardia, Pseudomonas, Rhodococcus,* yeast of the genera *Candida, Cryptococcus, Rhodotorula, Torulopsis,* and others, as well as the fungi genera Aspergillus, *Penicillium, Fusarium, Rhizopus.* Development of saprotrophic organisms is suppressed. Very high oil concentrations (*repression zone*) almost completely inhibit the soil biological activity.

Microbial community reaction to various petroleum product types is different from its reaction to oil. For example, soil contamination by engine oil and tar oil resulted in appearance of the two zones — homeostasis and repression. High pollutant concentrations corresponded to the homeostasis zone. In the repression zone, the development of both saprotrophic microorganisms and HOMs was inhibited.

Oil damaging effect on microbiota is comprised of light fractions' toxic effect on microorganisms and substantial increase in soil hydrophobic properties caused by heavy oil fractions. Severe soil pollution by oil is detrimental to its water and air regime and reduces availability of mineral nutrients for microorganisms causing complete suppression of the soil microbiota. The size of the homeostasis zone for the same pollutant varies depending on the soil and can be used as a measure of the soil tolerance to pollution.

The result of oil exposure varies for different groups of microorganisms with the same amount of oil because of the differences in metabolism and resistance to the pollutant. This is illustrated by comparing reactions of different microorganism groups in the soddy-podzolic soils (Albic Retisols) to pollution by salable oil in a field experiment (see Table 5.1).

Table 5.1

Reaction of different microorganism groups in the soddy-podzolic soils to oil contamination (dose 24 LL/m², 16 months after oil spill)
[Guzev et al., 1989; Artemyeva et al., 1988; Shtina, Nekrasova, 1988].

Group	Indicators	
of organisms	*Quantity*	*Species(sectional) diversity*
Actinomycetes	Dramatically decreases (15.8% of baseline value)	Decreases (50.0% of baseline value)
Fungi	Increases (137.0% of baseline value)	Decreases (55.6% of baseline value)
Yeast	Increases (108.5% of baseline value)	Decreases (83.8% of baseline value)
Algae	Dramatically decreases (5.5% of baseline value)	Dramatically decreases (8.7% of baseline value)
Microfauna	Dramatically decreases (7.1% of baseline value)	Dramatically decreases

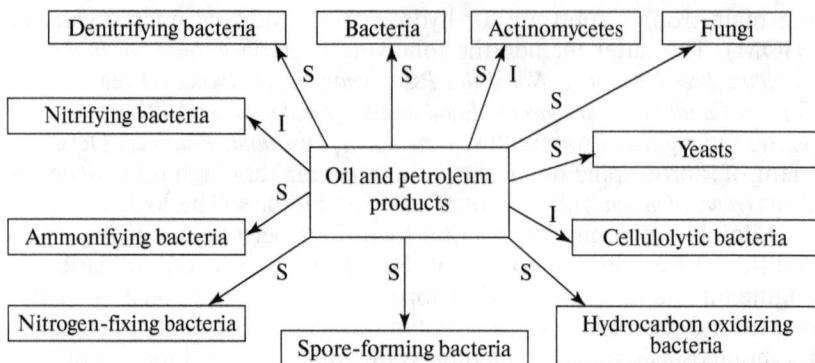

Fig. 5.7. The impacts of oil and petroleum products on the complex of soil microorganisms S — stimulation; I — inhibition [Ismailov, 1988, p. 49].

The controlled contamination level is the resistance zone for all groups of organisms being compared, as manifested by the decrease in their species diversity. However, the degree to which their composition changes is different. This reflects the varying resistance of considered organisms to oil pollution. Soil algae and microfauna are the least resistant to the impacts of oil. Their quantity and species diversity in contaminated soil dramatically decreases.

Based on simulation experiments and emergency oil spill studies in different natural zones, the reaction of different soil microbiota groups to oil pollution of soil can be represented as follows (see Fig. 5.7). Soil algae, microfauna, nitrifying, and cellulose-decomposing microorganisms have the least resistance to the impacts of oil.

Soil fungi

Although soil fungi (micromycetes) are tolerant toward oil impact, oil pollution affects their numbers and, particularly, species composition.

The quantity of fungi in oil-contaminated soils depends on oil concentration in the soil: it decreases in the "nucleus" of contamination aureole and increases in its peripheral parts. High population numbers in contaminated soils are typically produced by one or two oil-resistant species.

Even low (1–2%) oil content in the soil decreases the general species diversity and the number of dominating micromycete species, changes the species composition, and results in appearance of the species that are typical of the soils in more southern natural zones.

Qualitative indicators change is most pronounced in the "nuclei" of contamination aureoles. For example, in heavily contaminated high moors of Western Siberia (Surgut district), the species diversity falls to two species as compared to 7−10 species in uncontaminated soils. Domination of the *Penicillium* genus species gives way to domination of representatives of the *Trichoderma* (*T. aureoviride*) genus and of *Mycelia sterilia* group (TEREKHOVA, 2007). The most abundant and frequently occurring in these soils are *Penicillium thomii, P. spinulosum, P. lividum, P. funiculosum*. Fast-growing saccharolytic micromycetes are practically absent.

Pollution levels that result in microbiota transformation vary in different soils. For example, in Retisols of Western Siberia middle taiga, the impacts of oil on micromycetes are observed already at 2% content, while in high moors they are observed at 15% and more.

Soil algae

Change of soil algal communities under the impact of oil. Changes of algal communities are observed in hydrocarbon-contaminated soils: total number of species, species composition, structure, composition of dominating and distinctive species, ecological groups ratio, penetration depth of viable algae, and total number and biomass of algae.

The impacts of oil on algae depend primarily on its composition and concentration in the soil. Crude oil toxicity to algae is higher than that of salable oil. The reason is the toxic effect on algae that comes not only from the hydrocarbons but also from mineral salts if the soil is polluted by crude oil. If the initial pollutant load is exceptionally high (in various experiments, with oil dosage above 16 L/m^2, in emergency spill areas — in the "nuclei" of contamination aureoles), the soil can become temporarily "sterile." In case of very high levels of the soil contamination by oil, the destruction of algal community becomes catastrophic. Thus, even in half an hour time after 100 L/m^2 of oil (crude or salable) is introduced into soddy-gley soils (Umbric Gleysols), chloroplast deformation and discoloration are observed in the majority of algal cells. Within 7 days, algae population falls down to the values that are barely higher than the error of their quantitative determination method [DOROKHOVA, SOLNTSEVA, 2012].

The highest resistance to the impacts of oil is attributed to blue-green algae (cyanobacteria), colonial algae from the genera *Nostoc* and *Microcystis,* and filamentous algae from the genera *Leptolyngbya, Phormidium, Oscillatoria,* and *Lyngbya.* Yellow-green algae are the least resistant to the impacts of oil regardless of its composition: they disappear at minimum levels of soil pollution.

Weak soil pollution by oil (with low dosage in experiments and in the peripheral parts of contamination aureoles in oil fields) results in restructuring of algal communities that corresponds to the main directions of soil anthropogenic transformation. The species that are not resistant to the impacts of oil and soluble salts are replaced with resistant species (with the domination of blue-green algae). Halophilic species appear in the composition of algal communities. In case of reduction process intensification, the fraction of hydrophilic algae species in contaminated soils is increased substantially. Alkalization of contaminated soils results in a sharp decrease of the fraction of acidophilic and pH-indifferent species, alkaliphilic species predominate. Nevertheless, general species diversity of algae can be even higher than the level typical of algal communities in uncontaminated soils of adjacent territories.

Change of soil algal communities due to the impact of mineralized wastewater. Even short-term soil salinization (that occurs, for example, in the taiga zone) results in rapid (within the first year) transformation of algal communities in forest soils. Algocenoses are formed where blue-green filamentous algae from the *Oscillatoriales* order and certain diatoms species are dominant. Ochrophyta, typically, disappear or their diversity dramatically decreases. Halophilic and halotolerant species play an important role in algal communities of the soils exposed to heavy salinization. In terms of composition, they resemble algal communities of salinized arid soils unprecedented in the natural conditions of taiga. If higher plants are scattered or absent on the surface of salinized soils, algae form thick films that are similar to takyr crust. In such a case, algae become the main or the only ecosystem producers: their biomass amounts to $170\,g/m^2$ [SHTINA, NEKRASOVA, 1988].

Blue-green algae are the dominating group in undamaged soils of the forest-steppe and steppe zones (chernozems). Despite the differences between the varieties of agricultural use of chernozems, duration of their salinization and initial composition of algal communities, the general direction of their transformation due to the impact of mineralized wastewater is the same. In salinized chernozems, the diversity of blue-green algae is substantially decreased as compared to nonsaline soils, the diversity of green algae increases owing to ubiquist species, and the number of algae population falls sharply.

Soil invertebrates

Change of soil animal complex under the impact of oil. Soil animals in general are highly sensitive to soil pollution by oil. In a field experiment conducted in the forest-steppe zone of Tatarstan, even the smallest dose of salable oil ($6\,L/m^2$) resulted in a sharp decrease of both large invertebrates (mesofauna), and small arthropods population within 2 months from the moment oil was introduced. Soil pollution with salable oil at a dose of $24\,L/m^2$ killed the largest part of soil animals within days after contamination. Mass mortality of animals is caused by the oil direct toxic action, especially, of its light fractions. Population decreases mainly because of animal mortality in the top 10 centimeters layer of soil where oil content is the highest.

Heavy repression of soil animal numbers (both large invertebrates and small arthropods) was preserved in all soils for one year.

Earthworms are highly sensitive to soil pollution by oil. For example, in the forest-steppe zone, strong pollution of floodplain soil as a result of both a well accident and flooding with salable oil (24 L/m^2) resulted in their complete extinction in 5 years' time. Light oil fractions are the most toxic for earthworms. Manure worms are less sensitive to soil pollution by oil than typical soil earthworms. Among small arthropods, collembolans and oribatid mites have a particularly high sensitivity to soil pollution by oil. In the forest-steppe zone of Tatarstan, even the minimum initial oil load (6 L/m^2) on floodplain soils resulted in almost complete mortality of these soil fauna groups within a year. Besides a sharp decrease in numbers, in oil-contaminated soils their species composition is degraded: only occasional species of oribatid mites survive in the heavily polluted plough land areas (6–10% of oil). Species typical of human-transformed soils appear in their composition. The protozoa are the least sensitive to soil pollution by oil. Long-term research of their numbers and composition in the areas of emergency oil spills in various natural zones and in simulation field experiments identified their permanent presence in oil-contaminated soils. Reaction of protozoa to soil pollution by oil is similar to that of bacteria that consume the nitrogen of mineral and organic compounds [ARTEMYEVA, 1989; ARTEMYEVA ET AL., 1988]. On the population and organism level, oil impact is manifested by decreased survival rate, modified behavioral patterns of invertebrates, and morphological abnormalities, as demonstrated by the example of drosophilidae [PETUKHOVA, 2007]. This is caused by the oil mutagenic effect that is manifested even after short-term exposure of animals to the polluted environment. Change of soil animal complex under the impact of mineralized wastewater

In the oil producing areas of Tatarstan, even one-time salinization of Chernozem by sulfate-chloride-sodium effluent water resulted in a sharp decrease of both small and large invertebrate numbers. Total number of small invertebrates in the areas of wastewater spills decreased by almost five times. The number of oribatid mites decreased by 3 times under cropland, by 10 times under meadows, and by 30 times under forest; while the number of springtails decreased by 8 times under cropland and forest and by 20 times under meadows. Decrease in numbers mainly occurs due to animals' mortality in the upper (0–5 cm) soil layer. It is accompanied by depletion of species composition and by changes in the ratio of various small arthropod groups: the fraction of springtails and, especially, oribatid mites decreases. Soil salinization results in appearance of the species that are absent in undamaged soils.

Mesofauna numbers correlate with the general content of soluble salts in the soil and with Cl ion content due to high permeability of their covers to this ion. The numbers in salinized Chernozem under cropland reached 10.5 specimen/m^2, while in the nonsaline Chernozem, they were 18.5 specimen/m^2. The species that are typical of salt-affected soils were dominanin the complex of actively moving invertebrates (including ground beetles) that inhabit the upper soil layer.

Long-term pollution of forest-steppe Chernozems by oil field wastewater results in a still sharper decrease of pedobiont numbers. For example, on pasture, the numbers of protozoa decreased by 2.7 times; small invertebrates, by 10 times (the strongest decrease resulting from wastewater impact was observed in the number of oribatid mites and springtails), while large invertebrates completely died out.

Changes in the pedobiont complex are accompanied by weakening of the total biological activity in salinized soils: inhibition of ammonifiers and nitrate bacteria, as well as actinomycetes [ARTEMYEVA, 1989; ARTEMYEVA ET AL., 1988].

5.6. OIL AND FIELD WASTEWATER IMPACT ON HIGHER PLANTS

The impacts of oil production on soils

Morphological changes in plants. External signs of the plants troubled state in oil-contaminated areas are their morphological abnormalities: dwarf and gigantism, bent stems, twisted leaves and their change of color, necrosis, and tumors. Reduction of leaves assimilation surface and inhibition of reproductive organs formation is observed. Morphological changes in plants are accompanied by changes in their chemical composition resulting both from macro- and microelements ingress into soils with anthropogenic flows, and from changes in their migration conditions in oil field areas. Thus, in 3 years after an emergency spill of crude oil in the Komi Republic, the following background levels were exceeded in the polluted grass of hayfield and pasture land on the floodplains of the Kolva, Usa, and Pechora rivers: Hg (up to 4.1 times), Cu (up to 10.8 times), Cd (up to 2.8 times), Cr (up to 2.1–55.0 times).

Hydrocarbons consumption by plants. All plants, to a greater or lesser extent, are capable of consuming and metabolizing hydrocarbons. Gaseous hydrocarbons are consumed mainly by plant leaves. Roots consume aliphatic, aromatic, and polycyclic hydrocarbons; alcohols; and phenols.

Hydrocarbons that have been consumed by roots and leaves move into other organs with the transpiration stream and assimilants stream, penetrate into practically all tissues, and are accumulated in them. They are partially incorporated into the plant cell metabolite composition and partially released in the form of carbon dioxide gas. Some remain unmodified.

The ability to consume and transform exogenous hydrocarbons varies depending on the plant. A relatively high absorption capacity in terms of gaseous hydrocarbons is identified in English field maple (*Acer campestre L.*), Russian olive (*Elaeagnus angustifolia L.*), common almond (*Prunus dulcis (Mill.) D.A. Webb*), sweet cherry (*P. avium L.*), sour cherry (*P. cerasus L.*), Canadian poplar (*Populus canadensis Moench*), and common lilac (*Syringa vulgaris L.*) [UGREKHELIDZE, 1976].

Disturbance of the plant photosynthetic apparatus. Hydrocarbons "target" primarily the photosynthetic organs of plants with the mitochondria and chloroplasts being the most sensitive organelles of the cell. This results in disruption of photosynthesis and breathing in plants. In oil-contaminated areas, reduction of a and b chlorophylls in plant leaves and increase of a and b concentrations ratio are observed. This indicates that water-retaining capacity of leaves and plants resistance to external impact is reduced.

V.A. Veselovsky and V.S. Vshyvtsev [1998] in the southern taiga subzone (Permian Cisurals) studied the photosynthetic activity of Poaceae growing on oil-contaminated soil. The photosynthetic activity was studied using the method of plants long-lived afterglow kinetics (LLA) in field conditions. One year before site measurements, oil in amount of 8, 16, and 24 L/m^2 was introduced into soddy-podzolic soil (Albic Retisol). By the time of conducting the experiments, the top 5 cm layer of soil contained 3,000, 5,000, and 10,000 mg/kg of residual oil, correspondingly. The soils of baseline area contained 10 mg/kg of bituminous substances. Germinants of Hungarian bromegrass (*Bromus inermis Leyss.*) were studied in many replications in 2−3 weeks and in 1.5−2 months after sowing. Poaceae sown into highest-polluted soil sprung up but died soon. Their photosynthesis was virtually absent. In germinants that grew on the soil with the initial load of 16 L/m^2, the photosynthetic activity amounted to 50% of the baseline. Germinants with the lowest soil pollution in this experiment grew normally. Their photosynthesis was practically the same as baseline. Therefore, the impact of high doses of soil pollution by oil on photosynthesis is demonstrated quite clearly.

High resistance of the photosynthetic apparatus to the effect of benzene was identified in linden, maple, fir, poplar, spruce fir, walnut, plane tree, cypress, and ash [KVESITADZE ET AL., 2005].

Germinating capacity of seeds. Oil hydrocarbons decrease the germinating capacity of seeds. Hydrocarbons toxic effect on seeds is manifested at pollutant concentrations above 50 ml per kg of soil. In terms of their toxic effect on germinating capacity of lettuce seeds, (*Lactuca sativa L.*) lower alkanes (C_1-C_6) form the sequence of ethane < methane < propane, n-butane, n-pentane, n-hexane [UGREKHELIDZE, DURMISHIDZE, 1984].

Under otherwise equal conditions, oil toxic effect on the germinating capacity of seeds depends on soil pollution levels and plant species. Thus, at a 1% content of Fedorovskoye field oil in the soil, stimulation of summer wheat seeds germinating capacity was observed, although subsequent growth of young plants was retarded [SEDYKH ET AL., 2004]. At 5% oil content in the soil, wheat seeds germinating capacity sharply decreased, and young plant growth was inhibited. Germinating capacity stimulation of Siberian cedar and ash-leaved maple seeds was observed at oil content of 1.5−3.0%, while at 6% of oil content in the soil no consistent differences in the germinating capacity of these crops from the baseline values were observed.

Plants sensitivity and survival rate. Under natural conditions, soil pollution with crude oil produces the strongest impact on the vegetation cover, as it results not only in bituminization, but also in salinization of the soil. In the central parts of pollution, aureoles where pollutant concentration is the highest vegetation communities die out. In the peripheral parts, if the degree of contamination is low, a decrease in projective soil cover and species diversity is identified. Delay of the growing season start and its shortening, changes in the plant development rhythm to the extent of disappearance of certain phenophases is observed in many natural zones where oil pollution has occurred.

In terms of toxicity, aromatic hydrocarbons by far exceed alkanes. Toxicity of hydrocarbons of the ethylene series (ethylene, propylene, butylene) decreases proportionally to the increase of the chain length. Their high concentrations result in plant death.

Among herbage plants, the majority of meadow and forest species, as well as agricultural crops, are rather sensitive to oil.

Field simulation experiments aimed at studying the effect of oil from the Ust-Balyk field on poaceae growth and development in the taiga zone have demonstrated that oil pollution (both by crude and salable oil) resulted in lower germinating capacity of seeds and thinning of the grass stand [SHILOVA, 1988]. In terms of increasing sensitivity to oil under the conditions of taiga zone, poaceae form the following sequence: cock's foot > herd grass (*Agrostis alba L.*) > timothy grass (*Phleum pretense L.*) > English bluegrass > red fescue (*F. rubra L.*) > Hungarian bromegrass > upright brome (*Bromus erectus Huds.*) > American sloughgrass (*Beckmannia syzigachne (Steud.) Fernald*) > wild rye (*Clinelymus sibiricus L.*). For the leguminous, this sequence is as follows: large-leaved lupine (*Lupinus polyphyllus Lindl.*) > bird's foot trefoil (*Lotus corniculatus L.*) > alsike clover (*Trifolium hybridum L.*) > broad-leaved clover > white clover (*T. repens L.*).

Among tree species, coniferous trees are the most sensitive to soil pollution by oil.

In the northern taiga subzone of Western Siberia, in the areas of heavy or medium soil pollution by oil, the majority of adult trees in pine forest stands die in the first few years after an oil spill, while pine seedlings and young growth die completely. For the period of 10 years and more, seedling appearance is prevented by the bituminous crust that is formed on the soil surface. In the middle taiga, inside oil spills on high moors, the highest sensitivity to oil impact among the tree layer plants is exhibited by the common pine, while in forests it is the Siberian cedar (as compared to fir and spruce fir), and in the southern taiga it is the fir [SHVERGUNOVA, 2000].

In the southern taiga of the Kama region around Perm, tree species form the following sequence in terms of sensitivity to soil pollution with oil: pine > fir > aspen > birch [BUZMAKOV, KOSTYREV, 2003].

Impact of soil pollution by field wastewater on plants

Emergency discharges of mineralized wastewater produce a catastrophic effect on forest and meadow plant communities.

Significant impact of Cl-Na-Ca composition wastewater on vegetation at the time of the emergency was recorded at a distance of up to 5 meters from the spill. Almost complete death of plants that constitute the poaceous-forb community was observed in the central part of the wastewater impact sphere. Poaceae proved to be the most sensitive (turfy hair grass, bluegrass, and bentgrass). Hoary plantain (*Plantago media L.*) and silverweed (*Potentilla anserine L.*) proved to be more resistant to the impact of wastewater. Their individual specimens remained in a viable state in the polluted area. Very strong resistance to high mineralization of surface and soil and groundwater was exhibited by round-fruited rush (*Juncus compressus Jacq.*) that was growing in the depressions filled with mineralized water. In linden-birch fern-goutweed-sorrel forest with fir and spruce, almost complete death of ground cover was observed.

Among herbage plants, the following are relatively resistant to the effect of wastewater: hawk's beard (*Crepis sibirica L.*), great nettle (*Urtica dioica L.*), and white hellebore (*Veratrum lobellianum Bernh.*). However, these plants were smaller in size as compared to those that have grown in unpolluted areas. Tree species are highly sensitive to soil salinization: in the salt aureole "nucleus" undergrowth mortality and tree stand damage were recorded. This was especially pronounced in young trees.

In the peripheral part of the wastewater impact zone, no signs of plants inhibition were identified in the poaceous-forb community. Although, in the forest community partial death of ground cover, undergrowth death, and ee stand damage (yellowing of needles, leaves and last annual shoots) were observed [GLAZOVSKAYA, 1982].

CHAPTER 6
ENVIRONMENTAL IMPACTS
OF OIL AND PETROLEUM PRODUCTS
ON SURFACE AND GROUNDWATER

6.1. SOURCES OF SURFACE AND GROUNDWATER POLLUTION
IN OIL PRODUCTION AND PROCESSING AREAS

Petroleum anthropization creates a serious hazard due to pollution of main terrestrial fresh water sources—surface water and groundwater. Rivers, lakes, mires, ponds, and groundwater and water ecosystems in general are contaminated by hydrocarbons, oil, petroleum products, and field wastewater that contains salts of reservoir water from the deep layers of the lithosphere. This results from inflow of industrial, agricultural, and household effluents, as well as storm water run-off from polluted territories. Effluent water, regardless of its origin, always contains petroleum products, which is why it is one of the most common pollutants of terrestrial waters.

There are numerous specific sources of surface and groundwater pollution in petroleum production areas. Oil and petroleum products, field wastewater, drilling mud, and cementing slurry are released into surface water as a result of spills and chronic leaks from wells, pits, and other technical facilities.

Surface runoff and infiltration of wastewater with high salt concentrations through polluted soils are the major sources of continuous pollution of surface and groundwaters; in the areas of oil fields. Crossflow from deep reservoir water and oil into surface and groundwater via faulty wells or fracture cavities in the rock that surrounds the oil deposit also contribute to contamination of surface and groundwater objects. Besides oil, petroleum products and salts, surface and groundwater in oil fields may be contaminated by heavy metals, including mercury, and radioactive elements. The bulk of polluting substances is carried by flows of water into intermediate or final accumulation areas (lakes, reservoirs, seas). A substantial part of water-insoluble carbonaceous substances settles on the shores of watercourses and accumulates in bottom sediments contributing to secondary pollution of water.

The primary sources of surface and groundwater pollution in oil refineries and the territory around them are industrial effluent from production units and filtered water of effluent treatment plants. In terms of discharge to water bodies, oil refining and petrochemical industry ranks second among all other branches. Although oil refineries are typically located near large water basins, the volume of discharge and water consumption is so high that it disturbs their natural self-purification capacity [Kapustin, Gureyev, 2007].

Oil refinery effluent contains various toxic compounds, including propane, butane, ethylene, phenol, benzene, and other hydrocarbons. When this effluent is released into natural water, it negatively affects aquatic organisms and plants.

In terms of main indicators, effluent water composition only slightly varies depending on the oil refinery range of products.

The volume of discharged water per 1 ton of processed oil can amount to 70–100 m^3. However, its largest part (90–95%) is recirculated after going through appropriate treatment. Therefore, the actual quantity of effluent water in refineries is normally 1.6–3 m^3 per 1 ton of oil [Abrosimov, 1999].

Effluent water is removed from a refinery via two sewage systems. The first system is used to collect low-mineralized effluent and rainwater. After treatment, this effluent water is returned and re-used. Excess water (during rainstorms) is sent into emergency catch basins and discharged to a water body after treatment.

The second sewage system includes several (five to seven) networks that transfer effluent water from individual workshops and units. This water is highly mineralized, polluted by toxic substances, and is not used for recirculation. If required, it may undergo local treatment to remove specific contaminants.

6.2. SURFACE WATER POLLUTION

Oil and most hydrocarbons are hydrophobic. On average, just 1–3% (occasionally, up to 15%) of crude oil (mostly light hydrocarbons) dissolves in water, while between 10 and 40% of oil evaporates. Average saturation concentration is 26 g/m^3 for water bodies with weak water circulation and 122 g/m^3 for water flows. In the aquatic environment, oil can remain in different aggregate states for long periods:
- surface films (slicks);
- dissolved forms;
- emulsions ("oil in water" and "water in oil");

- suspended forms (residual fuel oil — oil aggregates floating on the surface and in the mass of water, oil fractions occluded on suspended solids);
- solid and viscous components settled on the bottom.

Currents and waves carry oil films to the shores where they are absorbed by coastal rocks and bottom sediments. Oil substance accumulation is particularly intense in the swamps. Oil, petroleum products, components of their modification and decomposition, and large quantities of heavy metals are accumulated in the peat mass that has a high sorption capacity. Peat mires act as a natural landscape-geochemical barrier and serve as traps for polluting substances. In the case of mires with flowing water, the concentrations of dissolved ingredients decreases downstream due to dilution, sorption, sedimentation, change of geochemical conditions, and consumption by the mire biocenoses.

6.3. GROUNDWATER POLLUTION

Groundwater pollution caused by economic activity results in changes of water quality making it partially or completely unusable. The main source of oil and petroleum products inflow to groundwater is oil-contaminated soil in the areas of oil production, transportation, and processing, as well as at oil depots and gas stations.

Hydrocarbons enter groundwater when precipitation water is filtered through polluted soils. Filtration of oil-containing effluent into the groundwater occurs quite often in the case of unconfined groundwater horizons. Significant quantities of oil hydrocarbons penetrating the groundwater change its hydrochemical and physicochemical properties.

The accumulation of hydrocarbons in focuses of groundwater pollution occurs in the following forms [GOLDBERG ET AL., 2001]:
- petroleum product lenses or zones of their film spreading that float on the surface of groundwater;
- dissolved and emulsified petroleum products in groundwater;
- heavy petroleum products near the lower periphery of the groundwater pollution zone.

A floating lens of petroleum products is the "nucleus" of the groundwater pollution zone. It is also connected with the gaseous "cap" of volatile hydrocarbons in the ground. Lenses mainly accumulate light hydrocarbon fractions (gasoline, kerosene, diesel fuel) with the highest penetrating capacity in the ground.

Total volume of dissolved petroleum products can amount to hundreds of kilograms per 1 km^2 in a secondary lens. Lens thickness can amount to 3 meters and more in its central part, while the lens area can be in the range of tens and, in certain cases, a few hundred hectares. Petroleum product reserves in floating lenses are tens to hundreds of thousands tons.

Hydrocarbons are partially absorbed by the rocks surrounding and underlying groundwater, where they can remain for many years. In this case, there is a risk of their subsequent release in the process of "washing off" by groundwater. Groundwater can turn into secondary focuses of natural environment pollution by hydrocarbons. This can create an environmental risk for the following natural and technical systems and facilities:

- reservoir water and bottom sediments in case of oil filtration on land and reservoir shelf;
- deeper water-bearing layers that are hydraulically linked to groundwater with the surface lens of petroleum products;
- supply intakes of surface and groundwater that happened to be on the pathway of an oil lens;
- soils;
- production facility basements and utility lines.

Intensity of hydrocarbon-polluted water movement in soils resulting in the groundwater pollution depends on the following factors:

(a) soil filtration properties and oil capacity;

(b) geological and geomorphic structure of the territory;

(c) groundwater table level;

(d) climatic characteristics of the territory;

(e) activity of microorganisms in the zone of aeration;

(g) physical and physicochemical properties of oil and petroleum products (their initial density, viscosity, boiling temperature, and water solubility).

Oil fields of the Apsheron Peninsula in Azerbaijan can serve as an example of groundwater pollution in oil production areas. On the Apsheron Peninsula, polluted groundwater has become a significant source of hydrocarbon pollution in the Caspian Sea offshore strip. Groundwater flow is directed from the central part of the peninsula towards the coastline. Out of 174.8 million m^3 of groundwater, around 98 million m^3 (over 56%) annually enter the Caspian Sea aquatic area both in the southern and in the northern parts of the peninsula [ISMAILOV, 2009].

Soil filtration properties in this area are highly variable depending on location. The hydraulic conductivity varies from a few tenths to 5–7 m/day and more in sands and sandy loam, and between 2 and 25 m/day in limestone. Oil capacity decreases in the following order: clay and loamy sediments > sandy loam and sandy sediments > calcareous rocks.

Presence of thick water-resisting or low-permeability rocks in the groundwater confining layer decelerates radial migration of pollutants, preventing contamination, but contributes to lateral movement of oil hydrocarbons. Radial migration of hydrocarbons is also retarded by interlayering of sediments of different textures (sand, loamy sand, and clay) in the zone of aeration.

If the groundwater level is located near the surface, the risk of its pollution by infiltrating hydrocarbon-containing water increases. Increased risk of groundwater pollution by oil exists in the northern coastal strip, the territory of Baku trough, and in Eastern Apsheron. Here, groundwater in sediments of the latest (Novokaspiiskaya) transgression of the Caspian Sea and in modern beach sediments is at the depth of 1.5−6.1 m; in sediments of the Khvalyn transgression, it is at the depth of 1−15 m. In recent years, the rise in the Caspian Sea level has led to the rise of the groundwater table in many areas of Apsheron. The risk of water pollution by hydrocarbons has also increased. A hazard now exists of groundwater transfer from the zone of aeration into the anaerobic zone, where the microbial decomposition of hydrocarbons is retarded, so they may be preserved in the sediments for a long time.

On the Apsheron Peninsula, the groundwater pollution by infiltrating water with hydrocarbons is less dangerous from May to September, when precipitation is almost absent and evaporation is intense. At the same time, climatic aridity and shallow groundwater level over a large part of the territory (20−30% of the peninsula's area) create the hazard of secondary soil pollution by hydrocarbons through evaporation from the polluted groundwater table and transpiration by plants.

Microorganisms affect the migration of hydrocarbons migration in the aeration zone. Firstly, in the process of hydrocarbons biodegradation by microorganisms, acid and alcohol-type compounds are formed and released into the environment. These compounds dissolve oil fractions and increase their mobility in the ground. Their increased mobility could also result from formation of organomineral compounds. Secondly, microorganisms form and release compounds with surfactant properties that facilitate the washing off of hydrophobic carbon compounds from the sediments and their participation in hydrocarbon migration flows.

Hydrocarbon pollution focuses are formed in groundwater due to the impact of anthropogenic factors. For example, in the central part of Apsheron, the concentration of light diesel fuel fractions in groundwater can amount to 7.5 mg/L, while that of heavy fuel is up to 2.3 mg/L. This indicates that about 140−150 and 40−41 t of the corresponding compounds are annually released into the groundwater on Apsheron. The level of hydrocarbon pollution of the groundwater in this region at a depth of 1.8–6.9 m exceeds allowable concentrations for fishery water bodies by dozens of times.

6.4. IMPACT OF WATER POLLUTION BY OIL ON FRESHWATER ORGANISMS

Some underground water, whose composition includes elevated amounts of naturally occurring hydrocarbons, may have therapeutic properties due to its high biological activity. Such water includes, for example, unique Naftusya water at the Truskavets resort (Cis-Carpathian region of Ukraine). This water contains from 1.5 to 30 mg/L of organic substances and has a typical petroleum odor. Naftusya is used to treat the gastrointestinal tract, kidneys, musculoskeletal system, and many other diseases.

Anthropogenic water pollution by products of oil extraction and processing results in completely different consequences. It constitutes a life hazard not only to individual organisms but also to entire ecosystems.

In water bodies that are polluted by oil and petroleum products, intense activity of hydrocarbon-oxidizing microorganisms results in high-rate consumption of free oxygen. When delivery of fresh oxygen is complicated (for example, in an ice-covered water body), scarcity of oxygen in water can occur resulting in fish kill.

Pollution of the aquatic environment by oil and petroleum products substantially weakens the functional viability of aquatic organisms.

Impact of oil-polluted water on aquatic organisms depends on the composition of oil and its fractions; it was studied in experiments with Anabaena variabilis cyanobacteria (blue-green algae) grown under different temperature conditions [VESELOVSKY, VSHIVTSEV, 1988].

The experiment was carried out using light low tar methane-naphthenic oil from the Samgori field, Georgia; its fractions—light (boiling temperatures 50–200 °C) and heavy (boiling temperatures 350–500 °C), and their mixture with the ratio similar to crude oil were applied. In terms of composition, the light fraction is similar to gasoline, and the heavy fraction, to residual fuel oil.

Cyanobacteria cultures grown at temperatures of 15 and 35 °C were used for the tests. The impact of oil and its fractions on cyanobacterial cells was evaluated by means of adding 0.5–8 mL/L of oil, 0.4 mL/L of light fraction, and 0.2 mL/L of heavy fraction to 11–12-day culture. Cell number and their photosynthetic activity were monitored by means of recording their fluorescence and long-lived afterglow (LLA) for the

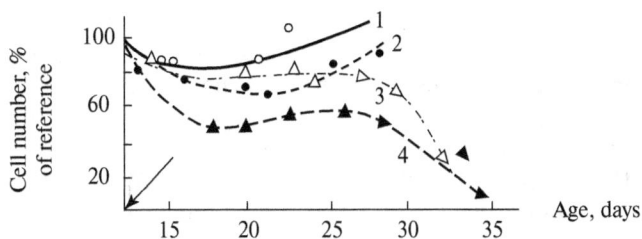

Fig. 6.1. Relative numbers of *Anabaena variabilis* cells grown at 35 °C (curves 1 and 2) and 15 °C (curve 3 and 4) after introducing oil into the medium (concentration 1 mL/L). (1, 3) *Anabaena variabilis* cell number determined using the fluorescence method; (2, 4) the same value determined using the LLA method (see Chapter 13). The arrow marks the time of oil introduction into the medium containing the culture [VESELOVSKY, VSHIVTSEV, 1998, P. 105].

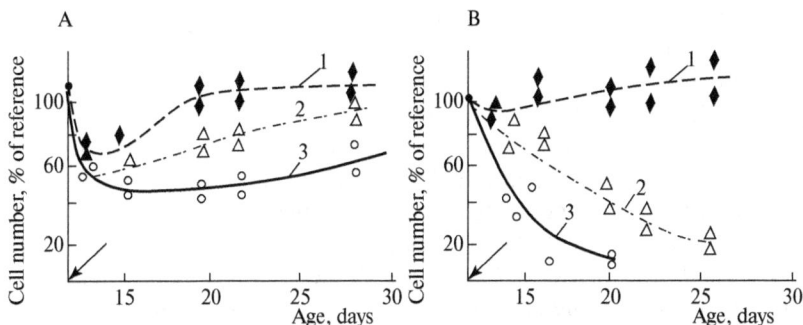

Fig. 6.2. Impact of light (0.4 mL/L, curve 1) and heavy (0.2 mL/L, curve 3) oil fractions and their mixture (curve 2) on relative number of *Anabaena variabilis* cells: (A) cultivated at 35 °C and (B) cultivated at 15 °C. Cell number was determined using the LLA method. The arrow marks the time of oil introduction into the medium containing the culture [VESELOVSKY, VSHIVTSEV, 1998, P. 105–106].

period of 10–35 days at a temperature of 22 °C. General level of cell fluorescence is proportional to their content of chlorophyll "a" and reflects the content for all cells. Living cell fluorescence was determined based on long-lived afterglow kinetic characteristics that reflect cell photosynthesis activity (dead cells are not engaged in photosynthesis and hence have no long-lived afterglow). Figures 6.1–6.2 illustrate the results of the experiment.

Fig. 6.1 illustrates the impacts of equal oil concentrations on blue-green algae cultures within 30–35 days after oil has been introduced into the medium. The cultures that have been grown under different temperatures react differently to the same contamination. In both cases, some cells are killed after pollution. Cell death happens quicker in the culture grown under cold conditions. After 30 days, the culture grown at 35 °C restored the initial cell number, while cells in the culture grown at 15 °C continued dying until no living cells remained, which happened on the 35th day.

Fig. 6.2 illustrates the results of cyanobacteria living cells exposure to the light and heavy fractions of the same oil in terms of their number. In case of pollution by light fraction, the number of living cells dramatically decreases in the first days. However, in two weeks a tendency for their number recovery is observed. Finally, in a month after light fraction was introduced into the medium, the number of living cells fully recovered in both cultures (see Fig. 6.2A).

When heavy fraction is introduced into the medium, cell death occurs much faster. In the culture grown at 35 °C, for almost one month the number of living cells remained at the level not exceeding 50% of baseline. Only on the 30th day, a weak trend towards restoration of their number was observed. In the culture grown at 15 °C, introducing the heavy fraction resulted in fast decline of living cell numbers with an irreversible degradation occurring on the 20th day (see Fig. 6.2.B).

Therefore, the effect of oil on organisms of even one species depends on the medium, in which it develops. Algae grown at higher temperatures proved to be more tolerant to pollution as compared to algae grown at relatively low temperatures.

The current point of view that the impact of light oil and petroleum products is more toxic than that of heavy oil proved to be not entirely correct. Light oil fractions and petroleum products like gasoline produce a fast short-term toxic effect similar to the influence of drugs that inhibit the activity of organisms. Organisms that survive the first "blow" can continue developing and participating in the ecosystem recovery. Light fractions evaporate, disperse, and degrade faster than others. Heavy oil fractions and petroleum products of residual fuel oil type are very tolerant to decomposition and contain a lot of toxic compounds. They produce a long-term toxic effect on living organisms resulting in complete degradation of ecosystems.

CHAPTER 7
ENVIRONMENTAL IMPACTS OF OIL
AND PETROLEUM PRODUCTS
ON THE MARINE ECOSYSTEMS

7.1. SOURCES OF MARINE ENVIRONMENT POLLUTION
BY OIL AND PETROLEUM PRODUCTS

Today, the problem of marine pollution resulting from petroleum anthropization has become a global one. Oil and refinery products rank among the top substances that pollute marine environments.

World Ocean, its intracontinental seas, and the largest landlocked water bodies have always been the field for the invasion of natural hydrocarbon flows and their accompanying chemical compounds. There are numerous natural sources of hydrocarbons in the ocean. However, not all natural phenomena that produce hydrocarbons flows in the marine environment cause the slightest damage to ecosystems. Economic activity of humans is the main hazard for World Ocean ecosystems.

Anthropogenic hydrocarbons are released into the marine environment in different ways:
- with the shore runoff of liquid and solid industrial and household waste by means of direct discharge from the shore, with river flows, with surface runoff, and with underground runoff;
- with atmospheric fallout;
- with the waste of marine vessels running on internal combustion engines;
- during excavation of various mineral resources on the seabed;
- because of chronic pollution and with emergency oil spills during petroleum exploration, production, and transportation.

According to different estimates, between 500 thousand and 8–10 million tons of anthropogenic oil and petroleum products are released into the World Ocean annually. It is not always possible to estimate the quantity of these flows due to the absence or inconsistency of actual data, as well as due to exceptional complexity and fast pace of natural processes on the sea-land and sea-atmosphere interfaces. In spite of large discrepancies between estimates, a trend can be discerned to gradual decrease of oil ingress into ocean from anthropogenic sources [NEMIROVSKAYA, 2013].

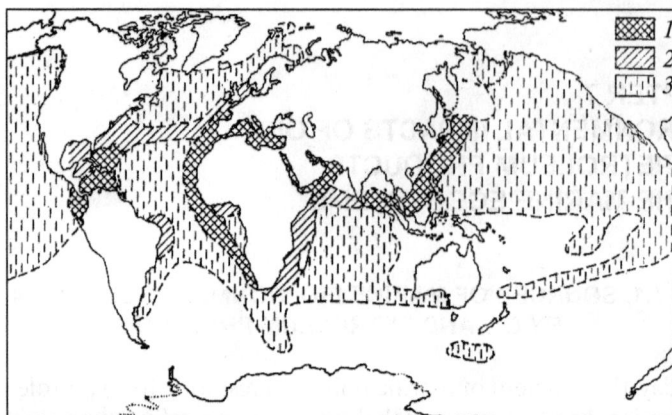

Fig. 7.1. Oil film on the World Ocean surface:
(1) found in 10—20% of observations, (2) found in 1—10% of observations,
and (3) found in 3—5% of observations [Nesterova, 1992, p. 146].

Emergency spills are the most dangerous among anthropogenic sources of marine environment pollution, although the volume of oil released into the marine environment due to accidents is relatively small compared to the total volume of received oil. Emergency spills from oil tankers result in thousands of tons of oil released into the sea, producing a destructive impact on marine ecosystems. The growing use of super-tankers and laying of oil pipelines on the seabed generates a high potential risk of marine environment pollution during oil and petroleum products transportation.

Fast-paced development of petroleum production industry and marine transportation results in pollution of large areas of the World Ocean and seas by oil hydrocarbons. Different quantities of oil films are found in all regions of the World Ocean (see Fig. 7.1). Most often, oil films occur near continental shorelines on sea shelves, where almost all the world oil-production industry is concentrated and marine transportation routes intersect. An oil film forms on the water surface as a result of oil and petroleum products being directly discharged into the sea. The largest quantities are discharged due to tanker and oil platform accidents and, to a lesser extent, due to oil and petroleum products runoff from shores. An oil film affects the processes of substance exchange between the ocean and the atmosphere, in particular, vapor exchange. Its presence is also a hazard to a normal living activity of the aquatic organisms that dwell in the upper layers of the water mass.

According to M.P. NESTEROVA [1992], oil in seawater is typically in the emulsified, colloid, or adsorbed state on suspended solids as resinous lumps. Oil emulsion is the predominant form of oil occurrence in seawater. Oil in the form of emulsion enters the sea with oil-containing effluent water, flushing water of oil tankers, and from other sources. Besides, spontaneous formation of emulsions is facilitated by the presence of high molecular weight components — resins and asphaltenes — in oil and heavy petroleum products.

Substantial impact of petroleum anthropization on the marine environment comes from petroleum exploration and production on the shelves of all continents. Continental shelves are environmentally highly sensitive water areas of the planet. Petroleum exploration operations performed here significantly affect the environment and human activities, including the fishing industry and the recreational industry.

Continuous negative physical impact on sea ecosystems is caused by geophysical exploration, operation of drilling rigs, pumps, and extraction of petroleum (Table 7.1).

Table 7.1

Factors of environmental impact on the sea at different stages of marine oil and gas field development [PATIN, 1997, P. 49]

Work stage	*Type of activity*	*Nature of the impact on the marine environment*
Geological and geophysical surveys	Seismic survey	Disturbance of the fishing industry and other users, impact on aquatic organisms and fish stock
	Prospective drilling	Seabed disturbance. Water areas isolation, process discharge, atmospheric emissions, emergencies
	Wells conservation and decommissioning	Disturbance of the fishing industry and other uses
Oil (gas)-field preparation and construction	Well testing	(See prospective drilling)
	Drilling platform installation, laying of pipelines, construction of onshore facilities, etc	Physical disturbance. liquid and solid waste discharge, disturbance of the fishing industry (platforms, pipeline) and other uses
Field operation	Drilling, process, transportation, and other operations	Process discharge during drilling and production, emergency spills and emissions, isolation of water areas, disturbance of the fishing industry and other uses
Production end and facilities decommissioning	Platforms and pipeline dismantling, well conservation, and other operations.	Discharge and isolation of water areas, disturbance of the fishing industry and other uses

The severity of oil spill consequences for the environment is normally estimated by the time required for recovery of polluted ecosystems. It is important for the adverse impact on an ecosystem not to exceed the difference between the period that separates two spills (or other adverse incidents) with the same consequences and the period required for the recovery of the resource after each of such spills [VINNEN, 1999].

7.2. IMPACT OF OIL POLLUTION ON MARINE ORGANISMS

Polluting substances negatively affect marine organisms and communities. At the same time, aquatic organisms are involved in transformation of these substances decomposing them into simple compounds and incorporating them into the general cycle of matter and energy in the World Ocean.

Plankton organisms

Phytoplankton. Different oil fractions and petroleum products affect algae differently. Aromatic hydrocarbons have a strong inhibiting effect on algae. The light fraction produces a strong but short-term toxic effect. Specimens that have not been killed immediately will recover their functions over time and will continue to develop. Heavy fractions produce a long-term negative impact on algae that results in irreversible damage of all cells. Hydrocarbons toxicity for algae also varies depending on the algae age. Thus, auxospores of diatom algae are less tolerant to the effects of oil as compared to adult cells [KUSTENKO, PODOLYAK, 1982].

Oil impact on sea algae is determined by its concentration in water, length of impact, and algae species sensitivity to oil hydrocarbons. Resistance/sensitivity to petroleum products and oil varies depending on the species of plankton and benthos-plankton algae. Most species of sea diatoms, dinophytes, yellow-green and green plankton algae are highly sensitive to crude oil. Oil concentrations of $0.01-1.0$ mL/L retard cell division and result in their death within the period of 5 days. Oil concentrations above 1 mL/L result in death of the majority of studied phytoplankton species. Zooplankton. Zooplankton that serves as fish and whales food is a component of marine ecosystems that is very sensitive to hydrocarbons impact. Its exposure to oil and petroleum products results in a toxic effect.

No relationship between zooplankton survival rate and pollutant type has been identified for the concentration of $0.001-0.1$ mL/L; no difference between species in terms of their tolerance toward seawater pollution by oil and petroleum products has been found (as opposed to phytoplankton). For some species, for example, Acartia, variation of tolerance towards the impacts of oil and petroleum products has been recorded for specimens of different genders. When grown in a medium with added oil pollutants at the rate of 0.001–0.05 mL/L, the number of living male organisms on the third–fifth day proved to be several times lower than the number of living female organisms.

Early stages of zooplankton development are particularly sensitive to sea water pollution by oil and petroleum products. Thus, at residual fuel oil concentrations in sea water of 0.01–0.001 mL/L, *Acartia clause* plankton larvae are killed on the fourth day, whereas adult specimens die on the sixth–eighth day, i.e., two times later.

Benthic organisms

Benthic algae. Adult macrophyte benthic algae are quite tolerant to seawater pollution by oil and petroleum products. For most species, under experimental conditions, the toxic effect of these pollutants appears at concentrations of 0.1–1/0 mL/L and higher. It is primarily manifested by a decrease in the intensity of photosynthesis, amount of pigments, and growth rate. For some algae species, low doses of oil and petroleum products were found to produce a stimulating effect on photosynthesis. The toxic impact of oil and petroleum products depends on the dose, duration of algal thalli contact with the pollutants and slightly varies in dependence on the form of oil occurrence in water (soluble components, emulsions, and films).

Benthic macroalgae include species both relatively sensitive to the effect of oil and petroleum products and tolerant to it. Apical parts of *Macrocystis sp.* brown algal thalli are sensitive to these pollutants. Even a short-term (3 hours) exposure to diesel fuel oil emulsion at a concentration of 1 mL/L results in reduction of their photosynthesis activity down to 55.6–56.6% of baseline values, in spite of subsequent staying in clean sea water for 5 days [MIRONOV, 1988]. The impact of residual fuel oil emulsion under the same conditions results in complete suppression of algal photosynthesis. *Polysiphonia elongata (Hudson) Spren.* red algae are sensitive to oil emulsion at a concentration of 1 mL/L. On the second day of exposure, its photosynthesis intensity is halved.

Under experimental conditions, the *Cystoseira barbata (Stack.)* C. Ag. brown algae proved to be tolerant to oil emulsion, as well as green algae *Enteromorpha intestinalis (L.)* Nees and *Ulva rigida C.Ag.* At a concentration of 1 mL/L, oil had practically no effect on these algae photosynthesis. Natural observations and experimental studies reveal that many fucus and laminaria algae are highly tolerant to the effect of oil. Thus, the rate of photosynthesis in *Fucus distichus L.* remained unchanged under exposure to oil emulsion at concentrations of 1–1,000 mL/L for the period of 4 days [SHIELS, 1973].

The mechanisms of sea macroalgae tolerance to oil hydrocarbons are not yet quite clear. One of them could possibly be explained by the effect of the hydrocarbon-oxidizing activity of symbiotic bacteria. This tolerance mechanism has been identified in *Fucus vesiculosus L.* Under laboratory conditions, a community of hydrocarbon-oxidizing microorganisms (HOM) consisting of *Pseudomonas fluorescens, Ochrobactrum anthropi, Rhodococcus fascians,* and dwelling on the surface of these algae thalli, reduced the content of diesel fuel by two orders of magnitude within three weeks [Voskoboynikov, Pugovkin, 2012]. It has been conclusively established that the NUMBER of HOMs on the surface of fucus thalli increases in case of seawater pollution by hydrocarbons.

The difference between hydrocarbon composition of macroalgae dwelling in clean and petroleum-polluted water, as well as the stimulating effect of low concentrations of hydrocarbons on photosynthesis of some species provide the grounds to hypothesize that the capability of certain algae to incorporate oil hydrocarbons into cellular metabolism is one of the mechanisms to resist them [Stepanyan, Voskoboinkov, 2006].

Under natural conditions, the diversity of macroalgae species in polluted water areas is depleted and their composition is changed. Long-living brown algae are replaced by green algae, the productivity of algae community, thalli size, and weight are reduced. At the same time, in areas with chronic hydrocarbon pollution of water, the adaptation of tolerant species of benthic macroalgae is observed. It is expressed in retaining their composition, quantity of photosynthesis pigments, and photosynthesis intensity at the levels typical of the specimens that dwell in clean water [Voskoboynikov et al., 2004].

Benthic animals. Adult benthic animals, similar to adult benthic macroalgae, are quite tolerant to seawater pollution by anthropogenic oil and petroleum products. Oil, residual fuel oil, diesel fuel oil, and kerosene produce a toxic effect on most groups of zoobenthos in the range of concentrations 0.1–1.0 mL/L and higher. The effect consists in motor activity inhibition, reducing a water filtration rate (for bivalve mollusks), feeding intensity, and survival rate. However, benthic animals also include both the groups of organisms that are characterized by strong tolerance to seawater and bottom sediments pollution by oil and petroleum products and organisms that are sensitive to it.

Natural observations and laboratory experiments reveal that certain bivalve mollusks — edible oysters (*Ostrea edulis L.*), Mediterranean mussels (*Mytilus gallopovincialis L.*), and cockles (*Cerastoderma lamarcki Reeve*) are highly tolerant to seawater pollution, particularly to water-soluble and emulsified oil components. It is established that water-soluble and emulsified oil components start having a toxic effect on Mediterranean mussels at concentrations above 20 mL/L, which is manifested in slower opening of their valves. Mussels stop actively filtrating water at pollutant concentrations of 50–100 mL/L [ALYAKRINSKAYA, 1966]. Sea urchins (*Strongulocentrotus purpuratus St.*) are also highly tolerant to the effect of oil and petroleum products. They are not killed at the 1 mL/L content of diesel fuel oil in water after 20–60 minutes exposure, although they lose the ability to cling to the substrate. After being placed in clean seawater, their motor activity recovers [NORTH ET AL., 1965].

Some benthic animals are relatively sensitive to the effect of oil and petroleum products. These, for example, include hermit crabs (*Pagurus bernhardus L.*) for whom kerosene is quite a toxic substance. After staying in water polluted by kerosene at the concentration of 0.01 mL/L for 2 hours, they were completely killed within 24 hours. High sensitivity to seawater pollution by oil is also characteristic for gammarus (*Gammarus olivii M.-Edv.*) and idothea (*Idothea baltica P.*) that are included into Malacostraca. Under laboratory experiment conditions, reduction of the adult idothea survival rate by 10–20% is observed at Malgobek oil concentrations in seawater of as low as 0.001–0.01 mL/L, after 20–30 days have elapsed. The adverse impact of oil on gammarus at the concentration of 0.01 mL/L occurs faster — in 10 days [MIRONOV, 1973].

Besides differences in species sensitivity to the considered pollutants, reaction of benthic animals to different petroleum products and oil compositions also varies to a greater or lesser extent. For example, for polychaete annelid worms *Nereis diversicolor Mül.* residual fuel oil is much less toxic than oil of Malgobek field. At concentrations of 1 ml/dm3 in the ground, mortality of these animals is observed in 40 and 5 days, correspondingly. For two gastropod mollusk species, *Bittium reticulatum* Da Costa and *Gibbula divaricata G.*, the opposite is true: residual fuel oil at 1 mL/L concentration is more toxic than oil from five fields, including the Malgobek field [MIRONOV, 1973]. The toxic effect of petroleum products and oil on the living activity of benthic animals is the stronger, the higher the pollutant concentration and time of exposure.

The young and larvae of benthic animals are more sensitive to seawater pollution by oil and petroleum products than adult specimens [MIRONOV, 1973]. For example, adult marbled crabs are highly tolerant to water pollution by residual fuel oil. At concentrations of 0.1–1.0 mL/L, they remain viable for 15 days. However, their larvae are killed at residual fuel oil, diesel fuel oil, and Malgobek crude concentrations of 0.001–0.01 mL/L between the 2nd and 9th day (depending on the dose).

Adult specimen of *Rhithropanopeus harrisiitridentatus* Maitland estuarine mud crab are highly tolerant to oil content in sea water, their lethal dose being equal to 60 mg/l of oil. Mortality of crab larvae is observed at oil concentrations that are two orders of magnitude less, i.e. 0.8–0.50 mg/l. Similar results have been revealed by the studies of living activity of the crabs that dwell off the shores of Alaska.

Mass mortality of marine benthic animals is observed at the time of emergency oil spills into the sea. For example, oil spilled during the accident of Amoco Cadiz tanker near the French shore in 1978 killed a large part of local echinoderms within the first week after the accident. In one year after the spill, benthic amphipods were still exposed to the adverse effect of oil that had penetrated into bottom sediments.

Thus, the outcome for benthic animals depends on the pollutant type and concentration, time of exposure, on the one hand, and organism species sensitivity and age, on the other hand.

Sea fish and marine mammals

Fish. Oil, petroleum products, and aromatic hydrocarbons have a strong impact on the living activity of fish. The first experimental studies of oil impact on fish in Russia were conducted at the end of the 19th century to handle the Volga River pollution issue. The study established that oil toxic effect on commercial fresh-water fish resulted from light saturated hydrocarbons and, in particular, from naphthenic acids present in its composition. Fish mortality was caused by light saturated hydrocarbons at concentrations of 1:5.000 — 1:3.000 and by naphthenic acids at concentrations that were smaller by 1–2 orders of magnitude.

Direct impact of polluting substances leads to various changes in physiological and biochemical processes of fish organisms, organ dysfunctions, disorders of fish higher nervous activity, and behavior. As early as the first days, or even hours, after water pollution by oil and petroleum products, the initial signs of fish intoxication appear lateral position, convulsions, loss of coordination. Later, these symptoms either deteriorate and result in fish mortality or vanish [KASUMYAN, 2001].

In case of a short-term exposure, reduction of fish feeding intensity is the only symptom, but is recovered already in 24 hours. A longer time of exposure to oil and petroleum products results in complete abandonment of food consumption by fish and to reduction of their viability.

Many species of fish, including highly valuable sturgeons and salmons, migrate from fresh water into sea (the young) and back (mature adults). If the routes of these fish migration go through shelf areas polluted by oil or petroleum products, migrating fish could be delayed or forced to change their migration direction, increase its length, etc. This could have a negative impact on fish reproduction, nutrition, and growth.

At the same time, adult fish specimens are much more tolerant to oil than other aquatic organisms. The most tolerant species are sturgeons, Caspian roach, and gobies. Fairly high tolerance to seawater pollution by oil and petroleum products is exhibited by first-year specimens of *Mugil saliens R.* mullet of the Mugilidae family, *Sargus annularis L.* gilthead, and Crenila brustinca L. East Atlantic peacock wrasse [Mironov, 1973]. They remain viable at oil (Malgobek field oil was used) and marine fuel concentrations of 0.1–0.25 mL/L within several days, while the mullet remains viable for several months.

Developing roe and young baby fish are affected by oil water pollution more than adult specimens. Under experimental conditions, the harmful effect of crude oil, diesel fuel oil, and residual fuel oil on roe of Black Sea turbot (*Psetta maeotica Pal.*), Black Sea anchovy (*Engraulis encrasicolus Lin.*), and ruff (*Gymnocephalus cernuus Lin.*) was observed even at low concentrations (10^{-1}–10^{-5} mL/L). It was manifested by substantial reduction in their survival rate, deceleration or acceleration of prolarva hatching, and significant increase in the number of abnormal (bent) nonviable prolarva. Similar to adult specimens, roe sensitivity to water pollution by oil and different petroleum products varies depending on the species. For example, residual fuel oil is the most toxic for Black Sea turbot and East Atlantic peacock wrasse, and diesel fuel oil is the most toxic for Black Sea anchovy [Mironov, 1973].

Water pollution by oil in fishing areas affects the quality of produced fish as a foodstuff. Even a short-term (6 hours) stay of salmon fish in seawater containing varying quantities of crude oil resulted in the fish acquiring an unpleasant odor and taste due to accumulation of aromatic petroleum hydrocarbons in its flesh. Undesirable organoleptic properties caused by water hydrocarbon pollution are the most pronounced in the fish that is characterized by an elevated level of fats and lipids in its tissues (sturgeon, salmons, etc.) The fish caught from contaminated areas also exhibit muscle tissue wetting and delamination, sharply intensified bioaccumulation of heavy metals (Pb, Cu, Zn, Mn, Cr, Ni, Cd), which content in certain cases increases by more than 1.000 times. As is well known, heavy metals have a strong toxic effect on many fish species.

Under otherwise equal conditions, the impacts of oil and petroleum products on fish significantly depends on the form of these pollutants occurrence. As compared to surface film, oil emulsion is more hazardous to fish because in this case, its toxic effect is accompanied by a mechanical effect — oil drops settle in the gills and disrupt breathing [Mironov, 1973].

Oil contamination impact also depends on a variety of accompanying factors: season of the year, water temperature, currents, weather conditions, etc., as well as on the specific oil properties.

Besides oil pollution, fish is affected by different geological survey operations: seismic survey, drilling, construction of utilities, etc. Physical impact on the marine environment results in reduction of fish nutritional reserves, its disorientation, especially during spawning, formation of man-made barriers, which the young fish concentrates around, and, eventually, mortality increase.

All of these changes result in disruption of ecosystem food chains, which is detrimental to the state of fish populations. Due to marine oil field development in the coastal area of Azerbaijan, it has lost its value for commercial fishing from cape Byandovan to Apsheron [KASYMOV, 2004] where not only animals, but also sea grass (*Zostera*) have disappeared.

Mammals. Oil pollution affects the state of the entire aquatic organism community, including mammals. It can disrupt the process of younglings development, reduce the reproductive ability of adult specimens, retard animal growth, affect puberty, and produce genetic deviations.

Birds that spend the largest part of their life on water are particularly sensitive to hydrocarbon pollution of the marine environment. Their feathering is soiled; oil ends up in their eyes causing irritation and in the intestinal tract resulting in intoxications. Severe accidents lead to catastrophic ecosystem changes, which eliminate the food reserves of sea birds and lead to their death. Bird eggs are the most sensitive to hydrocarbons; even small concentrations of these pollutants can kill the embryos.

Sea pollution by oil and petroleum products also affects polar bears, sea otters, seals, and fur seals. When the fur of these animals is contaminated, it becomes less efficient in holding heat, which leads to higher heat consumption for the animals. Oil causes eye irritation. If it enters the gastro-intestinal tracts, it can lead to hemorrhages, liver intoxication, kidney failure, and blood pressure variations. At the same time, sea mammals are capable of choosing their habitat area. For example, seals avoid the oil-contaminated island territories adjacent to the Apsheron [KASYMOV, 1994].

PART THREE

RESILIENCE OF NATURAL LANDSCAPES TO PETROLEUM ANTHROPIZATION AND LANDSCAPE REMEDIATION MECHANISMS

CHAPTER 8
THE CONCEPT OF LANDSCAPE RESILIENCE TO ANTHROPOGENIC IMPACTS

Three groups of environmental factors closely interact during ecosystem pollution by oil: (1) the heterogeneous composition and structural complexity of an ecosystem, (2) the diversity and variability of the external factors affecting the ecosystem (temperature, pressure, humidity), state of the atmosphere; state of the hydrosphere; etc.), and (3) the complexity and multicomponent nature of oil and petroleum products, as well as gases and field water.

Only if these three groups of factors are taken into account, can we adequately evaluate ecosystem resilience to oil pollution, the environmental consequences of oil pollution, and the ways to control them.

The first group of factors includes the composition, structure, and properties of landscape-geochemical system and its individual components—atmosphere, soils, surface water, groundwater, and biocenoses, or, case of marine systems, the water mass, bottom sediments, coastline, and aquatic organisms. These factors affect the pollution area, its substance composition, and the degree of contrast between the polluted area and the geochemical background.

The second group includes external factors that affect the rates of ecosystem degradation and ecosystem recovery processes. They determine the rates of abiogenic oxidation and dispersion of hydrocarbons and the activity of microorganisms. These processes develop differently in humid and arid areas, in the cold tundra, and in hot subtropical regions.

The third group of factors is the chemical composition of oil, petroleum products, and other polluting substances. "Oil" and "petroleum products" are a complex of substances completely different in terms of composition and properties, with different patterns of the natural decomposition process. There is very little in common between purification of the environment from condensate and heavy oil, motor fuel, and road bitumen. Besides, oil and petroleum products continuously change their composition and properties over time. Oil and petroleum products are accompanied by toxic associate substances (saline water, mercury, hydrogen sulfide, mercaptan, carcinogenic polycyclic aromatic hydrocarbons, radionuclides, etc.), that have a much more dangerous impact on the environment than oil itself.

The environment possesses the mechanisms that ensure adaptation and protection of ecosystems from pollution. This is a powerful factor for their self-preservation. Under different environmental conditions, natural mechanisms of ecosystem protection behave in different ways depending on resilience to a certain type of impact. Resilience of natural systems is a fundamental concept of geoecology.

According to M.A. Glazovskaya, resilience is defined as "the potential of a natural system to preserve its natural functional conditions." Resilience of natural systems to anthropogenic impacts is determined by: (1) the nature of anthropogenic impacts and (2) the properties of natural systems, their geochemical structure, and functional parameters [GLAZOVSKAYA, 1988]. On the one hand, the resilience of natural ecosystem depends on its ability to withstand anthropogenic impacts; on the other hand, it is specified by the capacity of this ecosystem to recover its properties after the anthropogenic impact is over and to return from the disturbed state to the normal state. The more resilient an ecosystem, the less time is needed to recover its functions.

M.A. GLAZOVSKAYA (1988) divides natural landscapes into the spatial groups that possess a similar level of geochemical resilience to anthropogenic disturbances of soils, surface and groundwater, atmosphere, and biocenoses across their entire area. Such groups have been termed *"technobiogeoms."*

Natural environmental factors play an important role in the resilience of technobiogeoms. Glazovskaya subdivides the set of indicators of the resilience of natural systems to anthropogenic impacts into three groups of factors responsible for (a) the removal and dispersion of the anthropogenic contaminants, (b) the intensity of their metabolism, and (c) the possibility and intensity of their binding in landscapes (Table 8.1).

N.P. Solntseva [1982] suggested assessing natural system resilience to anthropization using the "principle of anthropogenic impact compatibility with the natural system properties." Anthropogenic impact can be compatible with the direction of natural processes and intensify them. For example, the impact of mineralized reservoir water on landscapes in arid zones intensifies salinization of soils, rocks, and waters. Anthropogenic impact can be incompatible with the direction of natural processes. Thus, heavy soil contamination with saline reservoir water in the humid zone results in vegetation degradation, while acid soddy-podzolic soils (Glossic Retisols (Humic)) are transformed into anthropogenic Solonchaks.

Thus, resilience of natural systems to petroleum anthropization is a fundamental concept of oil and gas geoecology; it serves as the basis for assessment, protection, and remediation of the endangered environmental components.

Table 8.1

Factors that determine the resilience of natural systems to anthropization
[GLAZOVSKAYA, 1988]

Indicators	*Natural conditions*
I. Factors that determine the intensity of removal and dispersion of anthropization product	
Indicators of the dispersion and removal of anthropization products from the atmosphere	Precipitation (depending on the season);wind speed (depending on the season); runoff (depending on the season)
Indicators of the migration and removal of anthropization products from soils and running water bodies	Runoff/evaporation ratio; system position in the cascade system; soil texture
II. Factors that determine the intensity of metabolism of anthropization product	
Indicators of the energy of decomposition of pollutants	total solar radiation, kJ/year; accumulated temperatures above $0\,°C$; ultraviolet radiation; number of thunderstorms per year; rate of organic matter decomposition (plant falloff/forest litter ratio); intensity of photochemical reactions
III. Factors that determine the possibility and intensity of binding of anthropization products and their metabolites in landscapes	
Indicators of the intensity of binding of anthropization products binding in soils and their initial storage capacity	acid-base conditions; oxidation–reduction conditions; sorption capacity; humus amount; type of geochemical arenas (open/closed, strong/weak contrast); geochemical barriers; mineral composition of soils; initial pool of the elements that participate in anthropogenic flows; retention (retinization) of substances above the permafrost table
Indicators of local fallout of anthropization products from the atmosphere	number and duration of fogs per year; number and duration of windless periods per year

CHAPTER 9

RESILIENCE
AND SELF-PURIFICATION
OF THE ATMOSPHERE

9.1. FACTORS OF THE ATMOSPHERE
RESILIENCE

The atmosphere is a transit medium for anthropization products. The majority of gases, vapors, and aerosols released into air do not remain there for a long time. Molecules are dispersed in air and decomposed by means of chemical reactions, fall out onto the Earth's surface in the form of aerosols and precipitation. In the permanently polluted atmosphere, the concentration of anthropogenic substances is maintained by their release from the Earth's surface. The rates of decomposition of anthropization products in the atmosphere and natural purification of the air vary depending on the area. To assess the variability of decomposition of anthropization products in the atmosphere, M.A. GLAZOVSKAYA [2007] proposed using data on ultraviolet radiation intensity and the number of thunderstorm days as crucial indicators of the natural self-purification of the atmosphere. Ultraviolet radiation launches various photochemical reactions, including oxidation of gaseous anthropization products. During thunderstorms, anthropization products are quickly oxidized, and the pollutants are removed with precipitation. In the territory of the former USSR, ultraviolet radiation dosage varies from north to south between less than 100 and 800 W·h/m^2. The number of thunderstorm days per year is the smallest in tundra and forest-tundra (1–5) and in semi-deserts and deserts (5–10). The highest number of thunderstorm days per year (45–50) is in the mountainous areas of the Caucasus and Transcaucasia (Fig. 9.1).

Windless periods, when there is minimal air movement, reduce the rate of atmospheric self-purification. Probability of calms in the former Soviet Union varies in a broad range: between 10% and less on open sea coastlines to 50–60% in intermontane depressions and in continental areas of Central Siberia where the anticyclone predominates in winter (Fig. 9.2).

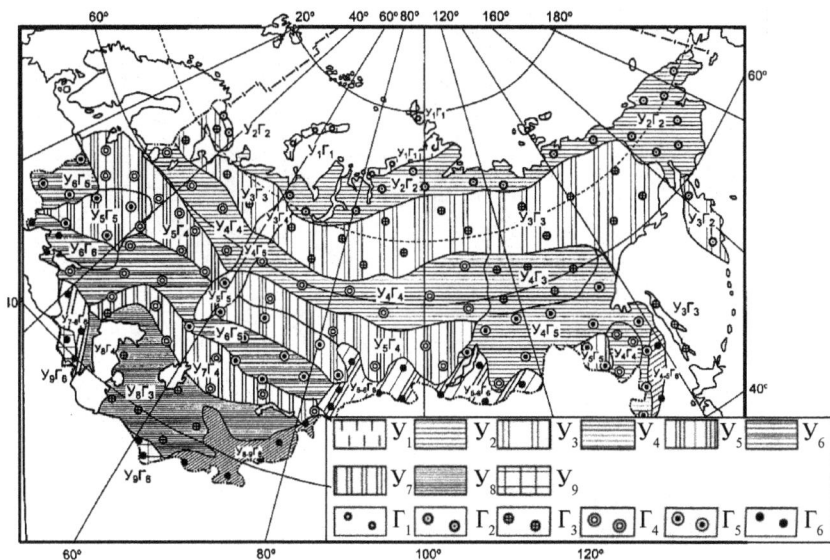

Fig. 9.1. Schematic zoning of Russia and neighboring countries in terms of potential intensity of decomposition of anthropization products in the atmosphere [Glazovskaya, 2007, p. 22].
Annual dose of total ultraviolet radiation, W h/m^2: Y_1 — less than 100; Y_2 — 100–200; Y_3 — 200–300; Y_4 — 300–400; Y_5 — 400–500; Y_6 — 500–600; Y_7 — 600–700; Y_8 — 700–800; Y_9 — more than 800.
Number of thunderstorm days per year: Γ_1 — less than 5; Γ_2 — 5–10; Γ_3 — 10–20; Γ_4 — 20–25; Γ_5 — 25–30; Γ_6 — 30–45.

Chemical processes that occur in the atmosphere are of great environmental importance. Earth atmosphere is a giant chemical reactor, in which solar energy induces oxidation and transformation of various chemical compounds, including anthropization products. These reactions result not only in decomposition of hydrocarbons and other pollutants but also in creation of even more toxic compounds. It is typical of atmospheric processes to include chain photochemical reactions that involve extremely powerful ionizing radiation. This results in formation of reactive electron-excited radicals — CH^{3-}, O^-, HO^-, HO^{2-}, H^-, etc. These radicals participate as intermediate products in the chain oxidation reactions of anthropogenic hydrocarbons, and nitrogen and sulfur oxides that are released into the atmosphere by the petroleum production complex.

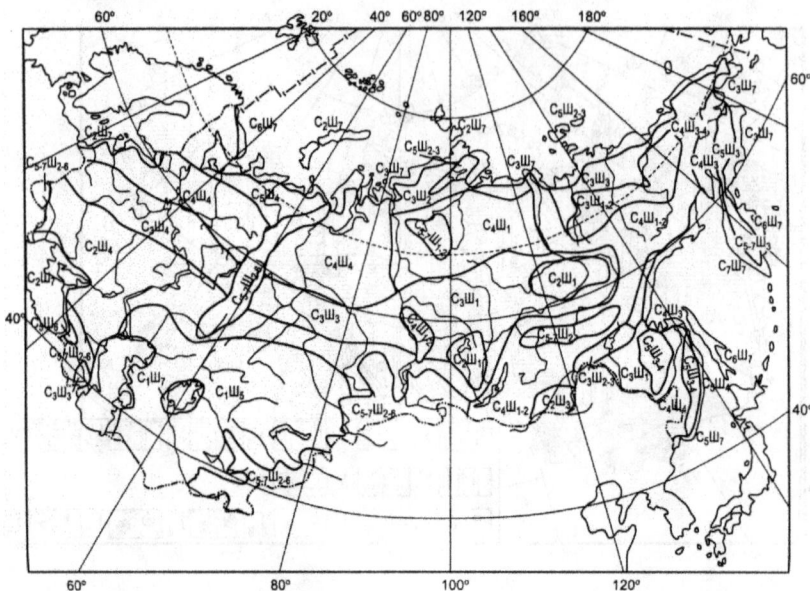

Fig. 9.2. Schematic zoning of Russia and neighboring countries in terms of potential intensity of self-purification from anthropization products via their dispersion with water and air flows [GLAZOVSKAYA, 2007, p. 326].
Annual water discharge (mm of water layer); $C_1 — 10$; $C_2 — 10–100$; $C_3 — 100–200$; $C_4 — 200–300$; $C_5 — 300–400$; $C_6 — 400–800$.
Probability of calms (%): $Ш_1$ — more than 60; $Ш_2 — 59–50$; $Ш_3 — 49–30$; $Ш_4 — 29–25$; $Ш_5 — 24–20$; $Ш_6 — 19–13$; $Ш_7$ — less than 12.

The presence of the active ozone molecule is vital for many atmospheric chemical processes. For example, hydroperoxide radical HO^{2-} forms in the troposphere due to the impact of sunlight as a result of ozone (O_3) and hydrogen peroxide (H_2O_2) decomposition and actively participates in NO oxidation. This radical plays a crucial role in formation of photochemical smog, a bluish haze that is a severe health hazard [BUKHGALTER ET AL., 2003]. Photochemical smog leads to increased morbidity and mortality of the population, vegetation damage, and metal corrosion at a faster rate.

Tropospheric oxidation is the primary way of the removal of methane (by more than 90%) from the atmosphere. An intermediate product of atmospheric methane oxidation is carbon oxide, subsequently transformed into dioxide, which, in turn, is involved in photosynthesis.

According to V.A. Isidorov [2001], methane oxidation proceeds differently in the clean atmosphere and in the atmosphere polluted by other compounds. In the "clean" atmosphere, reactions are initiated by the HO^- hydroxyl radical and proceed in several stages, producing CH^{3-} radical, water, formaldehyde, and carbon oxide. Ozone molecules are ultimately the main oxidizer:

$$CH_4 + 2O_3 \rightarrow CO_2 + 2H_2O + O_2.$$

Thus, two ozone molecules are consumed to completely oxidize one methane molecule.

In the polluted atmosphere, where large quantities of NO are present, NO_2 is added to the intermediate compounds. The lumped reaction is as follows:

$$CH_4 + 8O_2 \rightarrow CO_2 + 4O_3 + 2H_2O.$$

Thus, methane oxidation in the presence of nitrogen oxide results in accumulation of ozone in the atmosphere.

Oxidizing destruction of saturated hydrocarbons (alkanes) in the atmosphere begins with the removal of hydrogen atom through the interaction with HO^- and NO^{3-} radicals or atomic chlorine. Then, the oxygen molecule is added to the alkyl radical forming carbonyl compounds or alkylnitrites, compounds that are quite stable in the atmosphere.

Oxidation of aromatic hydrocarbons in the atmosphere proceeds by means of substitution of carbon atoms in the benzene nucleus by oxygen resulting in formation of phenol and cresols, or by means of ring opening and formation of derivative chain hydrocarbons. An important hydrocarbon sink, besides oxidation, is their consumption by plants or soil [Isidorov, 2001].

Therefore, the main factor of atmosphere resistance to anthropogenic impact is mechanical dispersion and chemical decomposition of anthropization products. Oxidation-reduction processes that proceed in the troposphere with the participation of excited molecules and radicals are important factors of atmosphere resistance. However, the products of anthropogenic substance oxidation include free radicals, oxidants that are harmful for health.

9.2. PROTECTION OF THE GROUND-LEVEL ATMOSPHERE. THE EARTH BACTERIAL FILTER

The Earth bacterial filter is a natural mechanism of protecting the atmosphere, a vital potential of the environmental resilience to hydrocarbon pollution. The bacterial filter in soils and the zone of aeration of

Phytocoenosis

Heterotrophic

SO_2
$(NH_4)_2SO_4$ S_{org} | | CO_2 C_{org} N_{org} | NO_2 | N_x NO_2 | N_2CO_2 C_{org} CO_2 C_{org} N_{org}

| Thionic bacteria | Hydrogen-reducing bacteria | Carboxido-bacteria | Methanol-oxidizing bacteria | 1st phase nitrifying bacteria | Ethane-, propane-, butane troph |

CH_3OH

NO_2^-

| | Methanol-oxidizing bacteria | 2nd phase nitrifying bacteria | Oil-oxidizing bacteria, fungi, actinomycetes, soil algae |

S_{recov}

CO CH_4 NH_3

H_2

S_0
Sulfide-forming bacteria
SO_4^{2-} | ? CH_3OH | NO_3^- NO_2^-

| Anaerobic bacteria | Methane-forming bacteria | Nitrate-reducing bacteria | C_nH_{2n+2} C_nH_m $n = 2-4$ |

CH_4

NO_2^-

| Denitrifying bacteria |

Lithosphere

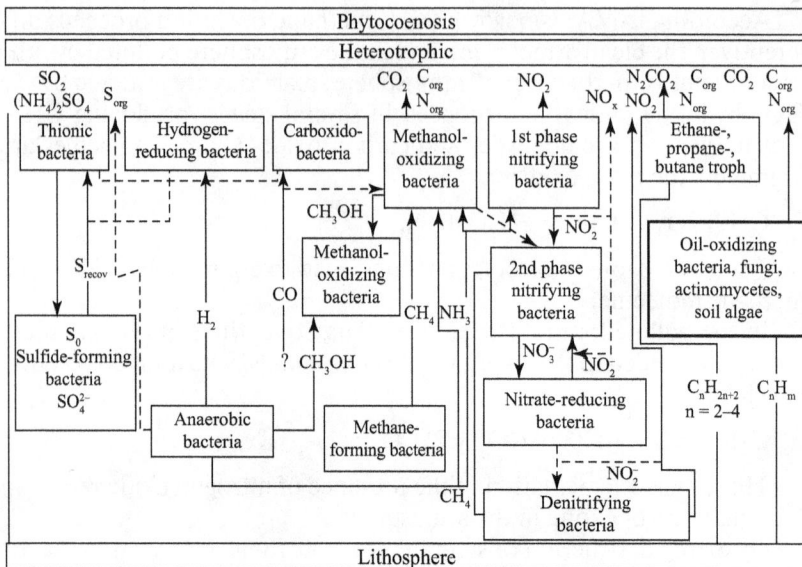

Fig. 9.3. Biocenosis of the pedosphere biological filter (according to G.A. Zavarzin. Cited from [OBORIN ET AL., 2004, P. 30]).

the lithosphere is a trophic natural system of microorganisms based on hydrocarbon-oxidizing bacteria (Fig. 9.3). The function of this screen is to control the inflow of hydrocarbons into the Earth's atmosphere by means of oxidizing a part of the hydrocarbons released into the atmosphere from the lithosphere and the landscape mantle of the Earth, including anthropogenic sources. The global biogeochemical value of the bacterial filter consists in maintaining the optimal conditions for existence of terrestrial ecosystems. The presence of a bacterial hydrocarbon-oxidizing filter in the zone of aeration of the Earth was established in the 1930s by a Russian geochemist G.A. Mogilevsky when studying traces of underground hydrocarbon accumulations in soils. A.A. Oborin made a valuable contribution to studying the Earth bacterial filter. He linked the bacterial processes of hydrocarbon oxidation in the zone of aeration near the Earth's surface to the functioning of underground biosphere [OBORIN ET AL., 2004].

Hydrocarbons, especially methane and its homologues, are the substrate and energy source for hydrocarbon-oxidizing bacteria that oxidize hydrocarbons into carbon dioxide and water. These bacteria stand in the beginning of the food chain. They assimilate the substrate, unusable for other representatives of the microbial community, and release exometabolites in the form of carbon dioxide, numerous amino acids, and the components required for heterotrophic growth — proteins, lipids, and carbohydrates. Existence and development of associated autotrophic and heterotrophic microflora is one of the mandatory conditions for hydrocarbon-oxidizing bacteria development. Bacterial oxidation of gaseous and liquid hydrocarbons is the essential function of bacterial filter performed by biocenoses. It is closely linked to the bacterial processes of nitrogen, sulfur, and hydrogen cycles.

The atmosphere is different from other environmental components because the boundaries of a contaminated area cannot be clearly identified. The contaminated space is highly variable. Pollution values can change within one day. On the other hand, the nature of petroleum production makes it almost impossible to prevent emission of pollutants into the atmosphere. Hydrocarbons, their combustion products, and geochemical associates are continuously released into the atmosphere in oil and gas fields, salable product parks, and oil refineries. Besides, there are numerous uncontrolled continuous flows of hazardous substances into the atmosphere.

The general ecological strategy to protect the atmosphere consists in keeping the total emissions volume as low as possible, and, on the other hand, in strict monitoring of the composition of controlled emissions and removing the most harmful substances from them. Conditions at the time of controlled atmospheric emissions must facilitate the highest possible dispersion of the substance. The criteria of atmospheric air quality should be in compliance with the specified hygienic standards for residential areas and industrial zones. As the concentrations of pollutants in the air are highly dynamic, maximum allowable concentrations are established for separate measurements (20-min-long air sampling) and for mean daily values per a given period (month, season, or year). The monitoring of air quality is complicated by the great number of substances to be monitored. The opinions concerning the way each of these substances affects living organisms and the general state of the atmosphere can change over time; hence, the standards can also change.

CHAPTER 10

RESILIENCE OF SOILS
AND VEGETATION
AND THEIR REMEDIATION

10.1. ENVIRONMENTAL FACTORS
OF RESILIENCE
AND SELF-PURIFICATION OF POLLUTED SOILS

The main criterion of soil resilience to oil pollution is its capacity for self-purification and recovery of initial functions. Biotic factors play the decisive role in the natural purification of soils and recovery of their fertility. The major role belongs to the complex of soil microorganisms that destroy organic compounds, including hydrocarbons. A set of conditions is required for their functioning: heat and moisture supplies, the presence of oxygen, and the presence of mineral nutrients. This set of conditions is ensured by the abiotic environment properties. They include mainly the climatic factors: temperature and humidity that determine the optimum soil biogenicity.

Water, air, temperature, and other regimes of soil have a strong impact on the abundance of certain groups of soil microorganisms, their functional activity, rate, and intensity of microbiological processes.

The factor of temperature (accumulated effective air temperatures and soil temperatures throughout the year) is one of the abiogenic factors that determine the intensity of microbiological soil purification processes after pollution by oil. Therefore, one of the most important steps in evaluating the ability of various landscapes to purify themselves from oil hydrocarbons is the assessment of the climate thermal potential.

When analyzing the availability of heat for soil microorganisms, one should take into account that the main role in soil processes in general and the processes of soil self-purification from oil hydrocarbons, in particular, belongs to their mesophilic group (optimum temperature for their growth is in the range of 20 °C to 42 °C).

As opposed to climate quality classification that is based on calculating the annual sum of daily air temperatures above 10 °, when evaluating the factors of soil natural purification from oil, it is feasible to consider the annual sum of daily temperatures above 15°, which is closer to the optimum sum of temperatures required for the functioning of soil mesophilic microorganisms. The climatic potential for soil natural purification can be determined using the following formula:

$$CP = \frac{\sum T > 15\,°C}{5475\,°C},$$

where CP is climatic potential; $\sum T > 15°$ is the annual sum of effective daily temperatures $> 15\,°C$, $5475\,°C$ is the annual sum of effective daily temperatures $> 15\,°C$ close to the optimum.

For zoning of territories, the climatic potential as an abiogenic natural purification factor can be expressed by the climatic coefficient.

In summer months, increasing temperature and decreasing humidity of the soil result in growing numbers of the microorganisms that are adapted to functioning under high soil temperatures, as has been illustrated by the example of Azerbaijan soils [MAMEDOV, ISMAILOV, 2006]. Many soils, including Chernozems and Kastanozems are subjected to winter freezing, which results in pronounced seasonality of the biological activity in these soils. In the areas of sierozemic and gray-brown desert soils, soil surface temperature throughout the year almost never falls below 0°, and microbiological processes are limited by the soil moisture.

Plants and microorganisms mainly consume nutrients from the soil solution. Therefore, the soil water content is of vital importance for the microorganisms living activity. Microbiological processes proceed with the highest intensity at the soil water content near 60% of the total soil water capacity. At this level of moistening, the soil is adequately provided with water and air. In a natural environment, soil water content is subjected to substantial fluctuations. In arid regions, it is not in summer that the soil microbiological processes are the most vigorous, but in spring and autumn, when the soil temperature and moisture are most favorable for development of microorganisms.

Although seasonal fluctuations of microorganism numbers in different soils could be quite substantial, these fluctuations remain within certain limits in each soil and characterize specificity of the microbiological and biochemical processes that occur in these soils. Therefore, averaged data on microorganism numbers within a long period of observations can characterize the biogenicity of different soil types. The total number of microbial population in the soil consistently grows from alpine and subalpine mountainous Meadow soils to steppe Chernozems and dry steppe Kastanozems. It is slightly lower in Sierozemic and Gray-Brown desert soils (Calcisols), in particular, in saline areas. The number of microorganisms in Brown Forest soils (Cambisols) is insignificant. In Yellow Ferrallitic soils (Lixisols), despite relatively benign temperature and moisture conditions, the number of microorganisms is low because of the high acidity.

Hydrocarbon pollution of soils in all natural zones results in intensified development of hydrocarbon-oxidizing microorganisms (HOMs).

Phytocenoses as a biogenic factor
of landscape self-purification

Not only microorganisms but also higher plants must be considered as a biogenic factor in of ecosystem self-purification from oil and oil products. This is explained by their potential ability to participate in degradation of hydrocarbons. It is known that plants absorb hydrocarbons through their roots and leaves. It is also known that they undergo various transformations in plant cells (Chapter 5).

Depending on the natural zone, the composition of plant communities, their productivity, and the surface-to-root biomass ratio are different, which determines their involvement in decomposing hydrocarbons.

The vegetation cover (particularly in the forest zone) is an active factor of hydrocarbons absorption from the ground layer of the atmosphere. This is an efficient aerosol filter, as, for example, the total leaf area in a broadleaved forest exceeds the area of crown projection on the soil by four–nine times. On average, 1 ha of middle-aged tree stand generates 4 tons of leaves (dry weight). If we assume that the average value for volatile hydrocarbons absorption by vegetation communities is 0.05–5 mg [UGREKHELIDZE, 1976], with the known biomass of vegetation we can calculate the total hydrocarbons absorption with respect to the contribution of each individual species. If we compare the community of microorganisms and higher plants in terms of their contribution to natural soil purification from oil and petroleum products, we will see that the microbiological factor plays the decisive role (Table 10.1). Owing to their small size, microorganisms have a larger specific surface and are characterized by the high intensity of matter exchange with the surrounding media. High growth and reproduction rates ensure that, under favorable conditions, the complex of soil microorganisms can take an active part in decomposition of oil hydrocarbons. The soil mesofauna is killed almost completely under the impact of soil pollution by oil and is not involved in the self-purification processes.

Table 10.1
**Reaction of the components of biocenosis to soil pollution by crude oil
and their role in the self-purification processes
[MAMEDOV, ISMAILOV, 2006, P. 68]**

Biocenotic components	Reaction to crude oil impact	Ability to use oil as a source of carbon	Role in landscape self-purification from oil
Mesofauna	Development almost completely suppressed	Absent	Not involved in natural purification
Higher plants	Development heavily suppressed	Very weak, mg/kg of biomass	Very weakly involved in natural purification
Microorganisms	Stimulation of certain groups growth	High for certain groups, 200–300 g/kg of biomass	Actively involved in natural purification

Cartographic assessment of soil resilience to hydrocarbon pollution

Potential ability of soils to remove hydrocarbons is assessed based on the factors of pollutant decomposition and dispersion. The method of forecasting soil resilience to hydrocarbon pollution was developed in the 2000s and implemented to compile maps of the relative resilience of soils in Russia and its regions [ECOLOGICAL ATLAS OF RUSSIA, 2002; NATIONAL ATLAS OF RUSSIA, 2007]. The method of soil zoning in terms of soil resilience to pollution by a wider range of organic substances (oil hydrocarbons, pesticides, dioxins, detergents) has been developed using the case of the territory of Azerbaijan Republic [MAMEDOV, ISMAILOV, 2006].

Assessment of soil resilience to hydrocarbon pollution is based on the cartographic assessment of the factors that facilitate or prevent natural purification of soils [GENNADIEV, PIKOVSKY, 2007].

The self-purification potential of soils depends on their capacity to decompose hydrocarbons via biological or physicochemical processes and to mechanically disperse pollutants and products of their metabolism. Both groups of factors are integrated in a matrix that is essentially the summarized expert assessment of natural purification conditions. The number of factor strength levels depends on the specified degree of assessment detail and on the availability of factual data.

The potential for natural purification is realized if the pollution level stays below the allowable limit for soils of this territory. These limits must be used as the basis for setting the regulations on soil pollution by oil and petroleum products.

The maps of soil resilience to hydrocarbon pollution are based on regional soil maps on the scale of 1:10,000,000 or larger.

Criteria of intensity of physico-chemical decomposition of hydrocarbons:
- oxidation-reduction regime of soils;
- sum of daily soil temperatures.

Criteria of intensity of biological decomposition of hydrocarbons:
- duration of the growing season;
- capillary moisture reserves;
- soil temperature regime.

Criteria of mechanical dispersion of hydrocarbons in the soil profile:
- annual precipitation;
- soil water regime.

Criteria of firm binding of hydrocarbons in the soil profile:
- thickness of organic and humus horizons;
- distribution of permafrost.

10.2. NATURAL DECOMPOSITION OF OIL AND PETROLEUM PRODUCTS IN SOILS

Abiotic and biotic factors of oil oxidation. When oil is extracted to the Earth's surface, it encounters a totally new set of conditions: from an anaerobic environment with very low rates of geochemical processes, it enters an aerated environment. In this environment, oil that consists of numerous individual compounds degrades very slowly. Oxidation processes of certain components are inhibited by others; transformation of individual compounds proceeds with production of oxidation-resistant forms. The main mechanism of various hydrocarbon classes oxidation under aerobic conditions similar to those on the Earth's surface is oxygen implantation into the molecule, and substitution of bonds with low dissociation energy (C-C, C-H) that are the most accessible for the oxidizer, with higher dissociation energy bonds (C-O, H-O).

Natural decomposition of crude oil in soils is facilitated by abiotic and biotic factors of its oxidation.

Natural abiotic factors of oil decomposition include (1) evaporation and dispersion and (2) physicochemical oxidation.

In the first weeks and months after contamination, oil mainly undergoes abiotic transformation processes in the soil: flow stabilization, partial dispersion, reduction of hydrocarbons concentration. This enables microorganisms to adapt, modify their functional structure, and begin actively oxidize hydrocarbons.

Oil content in soils is dramatically reduced in the first months after contamination, by 40–50% on average (this value is influenced by climatic conditions). Subsequent reduction of this value proceeds very slowly. The diagnostic indicators of residual oil change: the substance that was initially fully extracted by hexane is then substituted by the substance that is predominantly extracted by chloroform and other polar solvents. The fraction of substances that cannot be extracted from the soil by organic solvents, increases.

An important role in oil and petroleum products decomposition on the surface belongs to ultraviolet radiation that is capable of decomposing even the most resistant polycyclic aromatic hydrocarbons in a few hours. Photochemical processes do not occur inside the soil profile. Here, the primary role in hydrocarbon oxidation belongs to microorganisms. The main oxidizing agent under Earth's surface conditions is molecular oxygen. Hydrocarbon oxidation by molecular oxygen both in chemical and biological reactions occurs with the same mechanism and provides the same results.

Oil biodegradation processes are the most intense in the upper horizons of the soil profile where the highest oxygen content and the largest number of microorganisms are observed.

Hydrocarbons interaction with oxygen results in decreased level of free energy and can, therefore occur spontaneously if a sufficient number of oxidizing agents is available. At low temperatures, this process is extremely slow. One of the significant factors to accelerate the process is the catalyst activity. The most common and versatile catalysts in soils, water, and plants are various enzymes released by microorganisms, soil mesofauna, and higher plants.

Besides, there is a specific group of the hydrocarbon-oxidizing microorganisms that use certain hydrocarbons as a substrate for their growth.

Biodegradation of oil and petroleum products components. In a situation where the functional activity of soil flora and fauna has been almost completely suppressed, hydrocarbon-oxidizing microorganisms are at the forefront of petroleum products degradation in the soil. Soils and subsoil aerated horizons up to the groundwater level, as well as areas of neotectonic fracturing with elevated gas and water exchange, are the most favorable for the hydrocarbon-oxidizing microorganisms living activity.

Oil and petroleum products in the soil go down a kind of "bioconveyor" that stimulates their decomposition process and ecosystem purification [Ismailov, 1988].

This "conveyor" includes the following links:

(1) the hydrocarbon-oxidizing microorganisms that use certain hydrocarbons as a substrate for their growth;

(2) enzymes released by microorganisms, soil mesofauna, and higher plants;

(3) other groups of the microorganisms that utilize intermediate products of hydrocarbon oxidation and facilitate further intensification of their destruction.

In the zone of aeration, all oil components, including resins and asphaltenes, undergo transformations by natural microbial communities to various degrees and at a various rates.

The ability to oxidize hydrocarbons is widespread in the microbial world. 75 microorganism species capable of using aliphatic hydrocarbons and 25 arene-oxidizing species have been discovered. They are mainly representatives of the bacterial genera Pseudomonas, Mycobacterium, Achromobacter and of the yeast genera Candida and Torulopsis.

Representatives of the Pseudomonas genus have the widest capabilities and can use a wide range of aliphatic and aromatic hydrocarbons.

Hydrocarbon oxidation process may occur at different rates in the widest range of aeration, temperature, pH, and salinity. However, at present not a single representative of the micro world is known that is capable of consuming all hydrocarbon groups that constitute the basis of oil.

Stages of oil transformation in soils. Three most general stages of oil transformation in soils are identified:

(1) physico-chemical and, partially, microbiological decomposition of aliphatic hydrocarbons;

(2) microbiological decomposition of primarily low molecular weight structures of various classes, formation of new resinous substances; and

(3) transformation of high molecular weight compounds: resins, asphaltenes, polycyclic hydrocarbons.

The first stage of oil degradation in the soil takes between several and eighteen months, depending on natural conditions. It starts with the physico-chemical destruction of oil that is gradually supplemented by the microbiological factor. First of all, methane hydrocarbons (alkanes) are destroyed. The process rate depends on soil temperatures. For example, the content of this fraction decreased by the following numbers in a year-long experiment: in forest tundra by 34%, in middle taiga by 46%, in southern taiga by 55%. Simultaneously with alkane fraction reduction in residual oil, the relative content of resinous substances is growing.

The second stage of degradation takes around 4–5 years and is characterized by the predominant role of microbiological processes.

By the time of the third oil degradation stage, its composition has accumulated the most stable high molecular weight compounds and polycyclic structures, with the absolute content of the latter being lower.

The time required for complete oil transformation varies depending on the soil-climatic zone: between a few months and dozens of years.

Experiments on studying natural purification processes in oil-contaminated soils of different natural zones (tundra, middle and southern taiga, forest steppe, dry subtropics) were conducted in the 1970–80s under the supervision of M.A. Glazovskaya [1988$_2$]. Their results demonstrated that despite the observed differences in the rate of change for various hydrocarbon and oil fraction classes that depend on the soil-climatic conditions and oil composition, certain common features of its subsoil degradation could be identified. In all cases, oil content is reduced in the course of the physico-chemical and microbiological processes of its decomposition and mineralization, as well as transformation into insoluble or low-mobility forms. Processes intensity increases from north to south. For example, within a year in the arid zone around 50% of oil is transformed into various products of its microbiological metabolism that remain in their place. Within the same time, in soils of humid landscapes, the transformation of oil is less deep, and a large part of it migrates down the profile and/or is carried with surface or subsurface runoff beyond the original contamination areas (Fig. 10.1).

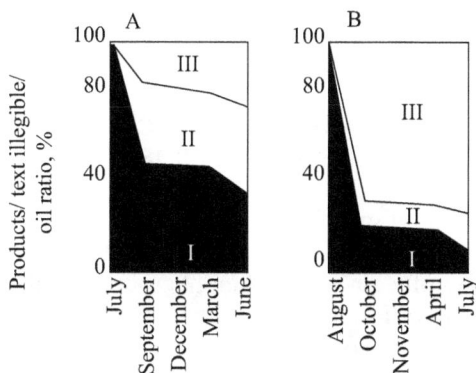

Fig. 10.1. Dynamics of oil degradation in soils:
A — gray-brown soil (Calcisol) of dry subtropics (insufficient moistening);
B — soddy-podzolic soil (Retisol) of southern taiga (excessive moistening).
I — residual oil; II — organic products of oil metabolism; III — oil mineralization products and the oil migrated from the contaminated area
[Glazovskaya, Pikovsky, 1980, p. 119].

Table 10.2.

The content of oil and its components in soils in one year after contamination (in % to original content) according to the data of simulation experiments (A1 horizon) [Pιкоvsкιy, 1993, p. 138]

Oil fraction	Arid zone, Apsheron Peninsula (Azerbaijan); gray-brown soil (Calcisol), heavy oil	Humid zone, Kama region; soddy-podzolic soil (Retisol), medium oil
Residual oil	35.0	7.5
Methane-naphthenic	15.1	0.35
Naphthenic-aromatic	12.5	6.8
Resins	94.0	36.0
Asphaltenes	50.0	12.5

The decrease in the residual oil content in soils is accompanied by changes in its chemical composition. With a general reduction of oil concentration in soils, the contents of its components change unevenly. Thus, the content of methane-naphthenic fraction decreases faster than that of other components. These hydrocarbons are easier to biodegrade. Besides, they have a higher solubility in water, which facilitates their removal beyond the initially contaminated areas. While the fraction of alkanes is reduced, the relative content of resinous substances in oil increases.

Methane-naphthenic and naphthenic-aromatic fractions are destroyed and removed at a much faster rate compared to other oil components (Table 10.2).

10.3. NATURAL PURIFICATION
OF SOIL BIOCENOSES

Self-purification of soils from oil and oil products and natural transformation of the latter "launch" the mechanisms of recovery of soil biocenoses. The rates of recovery of different groups of soil biota in oil-contaminated soils are different and depend on the resilience of particular groups to oil contamination.

Soil microorganisms

The recovery of microbial communities in oil-contaminated soils has been studied in field experiments conducted in different natural zones. These studies have revealed that the composition and number of different groups of microorganism in these soils vary in the course of oil transformation [OBORIN ET AL., 1988].

The first stage of oil transformation (1—1.5 years) is characterized by a high level of oil hydrocarbon content in the soil. After adaptation to the new environmental conditions (taking up the first few weeks), the microbial community undergoes a dramatic change. In this period, the leading role in its composition belongs to hydrocarbon-oxidizing microorganisms and the microorganisms linked to them via food chains that participate in oil hydrocarbons utilization. Typically, microbial communities include bacteria of the genera *Pseudomonas, Vibrio, Arthrobacter, Aeromonas, Acinetobacter; yeast of the genera Candida, Torulopsis, Rhodotorula;* fungi of the genera *Aspergillus, Penicillium, Fusarium, Trichoderma.* Hydrocarbon-oxidizing microorganisms (HOMs) quickly increase their numbers in oil-contaminated soils. In soddy-podzolic soils, their highest values have been recorded in 6 months after the contamination. At the end of the studied period, the number of HOMs in polluted soils decreases, however, in soils of middle and southern taiga it is still higher than the background level. Oil-contaminated soils contain higher numbers of microorganisms that participate in the nitrogen cycle — nitrogen fixers, ammonifiers, denitrifiers. At the same time, oil ingress into the soil typically results in severe suppression of nitrifying microorganisms development since the reduction conditions in soils are intensified and the content of soluble carbon compounds increases. A hypothesis has been put forward suggesting that reduction of nitrifiers' number in oil-contaminated soil facilitates the "compaction" of the nitrogen cycle: instead of nitrates, microorganisms receive nitrogen nutrition directly from ammonia that is released in the ammonification process [ISMAILOV, 2006].

In many oil-contaminated soils, the development of cellulose-digesting microorganisms is inhibited due to low nitrogen and nitrates content and an unfavorable water and air soil regime. The data on the effect of oil on actinomycetes are ambivalent: both an increase and reduction of their number have been observed in the first year after soil pollution by oil [ISMAILOV, 1988, KIREYEVA, 1995].

The second stage of oil degradation in the soil is characterized by a substantial reduction of residual oil quantity and a radical change of its composition. In this period (3—5 years), explosive growth of microorganisms' number is observed. Their most probable growth substrate is the compounds from long-chain alkanes group.

In peaty-gley soils of the tundra zone, oil biodegradation in this period is predominantly performed by heterotrophic bacteria; fungi and actinomycetes' activity is suppressed. In the taiga zone soils (podzols, swampy podzolic, podzolic and soddy-podzolic soils, peatlands), fungi, along with heterotrophic bacteria, play the dominating role in microbial communities. By the end of the studied period, the number of these microorganism groups falls down to the background values, although the number of actinomycetes grows substantially. In oil-contaminated gray forest soils, a stimulation of fungi, spore-forming bacteria, nitrogen-fixing microorganism growth was observed throughout the considered stage. Cellulose-digesting microorganisms and actinomycetes were inhibited. As opposed to unpolluted soils, fast-growing species were absent in the actinomycetes composition, while sporeless forms were predominant [Kireyeva, 1995].

The third stage of oil degradation is the longest. In the taiga zone, it is completed in 25 years after oil is released into soil. By the end of this period in southern taiga soils, the number of most soil microorganism groups in the soils exposed to oil contamination was identical to background level, except for hydrocarbon-oxidizing microorganisms which numbers were significantly higher than the background values [Oborin et al., 1988].

Soil algal communities

The rates of recovery of algae vary in dependence on the natural zone, but in all cases its beginning coincides with the start of the second stage of oil degradation in the soil, after most of the pollutant has been removed and the oil acute toxic effect on the soil biocenosis has finished. In the southern taiga subzone, the recovery of algae begins 1–2 years after the soil contamination. The pioneers to reclaim soils polluted by crude oil are small-celled green algae with the highest resistance to various extreme factors (including high solar irradiation and unstable substrate moisture content). These algae are also present in unpolluted soils (Yelshina, 1986; Shtina, Nekrasova, 1988, Zimonina, 1998). In the absence of higher plants in heavily polluted areas, these algae often produce an abundance that is up to 10 times higher than in uncontaminated soils of neighboring territories.

At the final stage of oil degradation (in 20 years after soil contamination) in the taiga zone, the content of organic carbon in upper soil horizons is identical to background, although other soil indicators are still far from the background values [Solntseva, 1998]. Therefore, in 20 years after heavy contamination of soddy-podzolic soils (Glossic Retisols) by crude oil no recovery of algae groups is observed. Species diversity is less than 50% of the initial level in unpolluted soils, many species of ochrophyta and green algae typical of soddy-podzolic soils with forest vegetation are absent. The complex of dominating species is changed dramatically: it includes representatives of blue-green and diatom algae (as opposed to unpolluted soddy-podzolic soils under forest where green algae dominate).

In the forest-steppe zone, blue-green algae are the first to reclaim the oil-contaminated soils.

Salable oil is less toxic for algae. Simulation of soil pollution process by salable oil using the "dose-time-effect" principle allowed to evaluate the rate of algae groups' recovery in the soils of different natural zones [YELSHINA, 1986; SHTINA, NEKRASOVA, 1988, ZIMONINA, 1998].

In the taiga zone, with the same initial load (25 L/m^2) the rate of algae groups' recovery is the highest in soddy-podzolic soils (Glossic Retisols) of southern taiga. Here, as early as in the third year after application of Surgut field oil, the general species diversity of active algae flora was 53.8% of the original value. However, the development of blue-green algae and, in particular, ochrophyta in the contaminated soil remained heavily inhibited.

At the same time, in illuvial-humus podzols (Albic Carbic Podzols) formed in the middle taiga subzone, contaminated with the Surgut field oil, only initial algae groups were formed consisting of four small-celled green algae species. The species diversity of such groups amounted to just 14.8% of the original value.

The lowest rate of algae community recovery across the experiment was recorded in the forest tundra, in the peaty-podzolic-gleyic soils (Gleyic Histic Retisols) contaminated by the Vozeysk field oil. With a relatively low level of initial load (12 l/m^2), oil pollution kept suppressing algae development for a long time. In 6 years after oil application, only initial groups of 8 green algae species were formed in these soils (corresponding to 25% of original species diversity).

Comparison of algological studies data with residual oil content in the soils of contaminated areas demonstrated that with the same initial load, the rate of soil algae groups' recovery was primarily determined by the rate of upper soil horizons natural purification from oil hydrocarbons [DOROKHOVA, SOLNTSEVA, 2012].

Soil invertebrates

Soil pollution by oil produces a lasting inhibiting effect on soil animals. Natural recovery of the soil pedobiont complex is very slow (ARTEMYEVA ET AL., 1988; ARTEMYEVA, 1989). Simulation of soil pollution by oil in different natural zones (saleable oil was used in the experiment, dose 24 l/m^2) revealed that an acute repression of pedobiont number in the soils of forest tundra, middle taiga, and dry subtropics is maintained for 3 years after application of oil. Within the 3 years of the experiment duration, recovery of the soil animal complex began only in southern taiga and forest steppe. However, there were distinguishing features in each of these natural zones.

In southern taiga, in the forest soddy-podzolic soils (Glossic Retisols (Humic)), the number of large invertebrates was restored in 1 year after an oil spill. The number of earthworms (one of the predominant animal groups in the mesofauna of these soils) increased thanks to the *Allolobophora diplotetrotheca* ground litter species that is dominating in unpolluted soils of neighboring territories. Typical soil species did not develop. In 3 years since the experiment beginning, the number of earthworms and large invertebrates in general in contaminated soils was higher than the background level. Within this period, small invertebrates practically did not develop. The restoration of the number of mites began earlier than that of other groups of invertebrates, but their trophic composition was unstable and significantly differed from their composition in the unpolluted soil.

An extremely high content of oil in the soil (often observed in the emergency spill areas) inhibits mesofauna development for a long time. The number of earthworms in heavily contaminated soils remained lower than in unpolluted soils of neighboring territories even in 15 years after contamination, especially in the litter horizon. By this time, the complex of small arthropods was already formed.

In the forest steppe zone, in chernozems under dry meadows the recovery of pedobiont complex in 3 years after application of oil began with small arthropods, one of the most numerous groups of soil animals in unpolluted soils. In 3−5 years' time after oil application, pedobionts were mainly represented by pioneering groups consisting of collembolans and mites (mole, acarid, and thread-footed mites). The number of collembolans in forest steppe chernozems recovered thanks to surface eurybiontic species. However, in three years under field experiment conditions, their species diversity amounted to just 37.5% (on plowland) and 61.1% (on meadow) of the background level. A noticeable amount of typical saprobes, oribatid mites, in the cultivated soil was observed only in 9−10 years after contamination.

In the areas of emergency oil spills, the rate of pedobiont complex recovery in cultivated and meadow soils was different. In 8−10 years, the complex of small and large pedobionts in cultivated chernozems was near the background level in terms of quantity indicators, although the diversity of soil invertebrate species was still low.

In Chernozems, under meadow vegetation where in original soils earthworms are represented by the typical soil species of Eisenia uralensis Mal., in the industrially polluted areas the populations of this species became completely extinct in 5 years. During the same period, the complex of small arthropods and the number of nematodes fully recovered.

Therefore, there are distinguishing features of the pedobiont complex recovery in oil-contaminated soils depending on the natural zone. It occurs after the process of soil natural purification from oil, transfor-

130

mation of its composition, and recovery of the vegetation cover. To some extent, the rate of this process is influenced by the type of land use.

Recovery of the pedobiont complex in soils exposed to salinization by mineralized oil field water is also slow. In forest-steppe Chernozems on plowland (Tatarstan), the number of pedobionts even after one-time salinization even after 9 years was substantially lower than in the soils of neighboring territories. After multiple salinization on pasture, only sporadic mesofauna representatives were found after 10 years, while the microfauna number was 2.6 times less than in unsalinized chernozems.

Thus, the way soil biota recovery proceeds is closely linked to oil transformation in soils, as the different stages of this process are characterized by the certain distinguishing features of bituminous substances composition and quantity. This explains the staged character of transformation of biocenosis in oil-contaminated soil.

10.4. NATURAL PURIFICATION OF HIGHER VEGETATION

Areas of mechanically disturbed soil and vegetation cover

After the stage of field construction is completed, the mechanical load is significantly reduced in many areas, which results in the beginning of vegetation cover recovery. The nature and intensity of recovery processes depend on the damaged surface area and the recovery potential of vegetation communities in neighboring territories.

The recovery is the faster, the more moisture is available in the area and the higher the role of herbaceous plants in the phytocenosis. For example, in western Yamal the highest recovery potential is typical of low-moor swamps and wet grass-mossy tundras. In the drained territory, even if less than 50% of the area has been mechanically disrupted and even if the complete vegetation canopy has formed relatively fast, the original phytocenosis recovers slowly. In case of large-scale mechanical disturbance (more than 50% of the surface is damaged), substantial changes in the thermal regime and intense relief transformation result in the lengthy process of secondary community formation and their significant difference from native communities in terms of species diversity and composition. For example, in Yamal after relief stabilization tundra communities are replaced by meadows consisting of grasses, sedges, and cotton grass [MAGOMEDOVA, MOROZOVA, 1997]. Due to the length of this stage, the problem of decreasing phytocenotic diversity in the tundra zone is being discussed. In Western Siberia, the number of angiosperm plant species on unpolluted anthropogenic sands is four times lower than in undamaged tundra areas, while the fraction of boreal species is noticeably higher [CHALYSHEVA, 1993].

Oil-contaminated areas

Vegetation recovery in contaminated areas is normally slow and has distinguishing features in different natural zones. Natural recovery of tundra vegetation after emergency oil spills in the European part of Russia requires, by some estimates, at least dozens of years [YUDAKHIN ET AL., 2002]. On emergency oil spill areas in the European tundras and forest tundra (Nenets Autonomous District and the Komi Republic) at high contamination levels (60–80 L/m^2) when the film of non-decomposed oil is preserved on the soil surface, the vegetation is completely absent [CHALYSHEVA, 1993]. This may be observed in a few years' time from the moment of contamination. Sporadic specimen of Scheuchzer's cotton grass (*Eriophorum scheuchzeri Hoppe*) and water horsetail (*Equisetum heleocharis Ehrh.*) are found only on small tussocks. Vegetation recovery from heavy pollution is more intense in swamp areas where at the initial healing stage (up to 7 years after contamination) 15 species of herbaceous plants can be found (average number of species — 4 per 100 m^2), while the total projective cover is between 2–3% and 30%. The dominant role in these groups belongs to bean trefoil (*Menyanthes trifoliate Dumort.*), Russet cottongrass (*Eriophorum russeolum Fries.*), water sedge (*Carex aquatilis Wahlenb.*), and water horsetail. Plant stimulation is observed quite often: their overall size or the size of certain organs become larger. Within the first seven years after oil pollution, shrubs and sub-shrubs are completely absent from plant aggregations being formed, as is the young growth of trees. Rhizomatous and creeping-rooted species play the leading role.

In the north of Western Siberia, polluted areas of ridge-pool swamps were almost completely taken over by sphagnum moss, sedge, and sub-shrubs in 14 years after the emergency oil spill [NOVIKOV, 1984]. In the middle taiga of Western Siberia, natural recovery of vegetation after an emergency oil spill on upland pine swamps begins only on the fourth year: pines are partially needled, swamp sub-shrubs offshoots appear, and repopulation by cotton grass, reed grass, rush, and sedge takes place [MAKOSVKY, 1988]. In the southern taiga of the Kama region near Perm, after soil pollution by crude oil only grassy–forb and willow shrub communities with young growths of trees were formed in the place of the linden goutweed-woodruff forest [GLAZOVSKAYA, 1982]. Full phytocenosis recovery in the taiga zone requires dozens of years. Sometimes, their changes are irreversible. In the forest steppe zone of Bashkiria, natural recovery of vegetation on soils heavily contaminated by oil begins with forming pioneer aggregates from oil-tolerant species of weed plants:

knotgrass, barnyard grass, couch grass, Canadian thistle, lamb's quarters, trailing bindweed, etc. [MINIBAYEV ET AL., 1986]. Natural recovery of phytocenoses in Bashkiria forest steppe requires 10 years at the level of soil pollution by oil up to 10%, while in case of heavy contamination it extends to several decades.

Areas contaminated by drilling waste

Natural recovery of vegetation at drilling sites has its specificity depending on the natural zone. Its rate and character depend on the degree of soil mechanical disturbance. contamination by phytotoxic agents, and moistening of the territory.

In the tundra and forest tundra zones (Nenets Autonomous District, Komi Republic), the leading role in plant communities at the initial stage of natural healing of drilling sites (3—4 years after drilling work is finished) belongs to forbs and grasses: alpine meadow grass (*Poa alpina L.*), annual meadow grass (*P. annua L.*), lapland reed grass (*Calamagrostis lapponicum (Schrank) Kunth.*), field horsetail, wood horsetail (*Equisetum silvaticum L.*), and sedges in small depressions. The area covered by plants is not more than 20% of the total area disturbed during drilling works. The number of vascular plant and bryophyte species per 100 m² is no more than seven [CHALYSHEVA, 1993]. The second stage of vegetation recovery (5—8 years after drilling work is finished) is characterized by a higher complexity of the vegetation cover structure, larger area covered by vegetation, and greater plant diversity. Grasses become predominant in vegetation communities — turfy hair grass, smooth meadow grass, annual meadow grass, narrow reed grass (*Calamagrostis neglecta (Ehrh.) Gaerth.*). The average number of species per 100 m² increases to 20. The third stage of vegetation recovery begins in 9—12 years after drilling work is finished. It is characterized by shrub and sub-shrub appearing in the composition of vegetation. Willows are the quickest to recover (*Salix lapponum L., S. glauca L., S. phylicifolia L., S. lanata L., S. rosmarinifolia*): their projective cover is 10% on average. The dwarf birch (*Betula nana L.*) plays a secondary role in formation of a shrub layer on drilling sites. Its projective cover does not exceed 3%. The projective cover of subshrubs — bog rosemary (*Andromeda polifolia L.*), wild rosemary (*Ledum palustre L.*), creeping rosemary (*L. decumbens Ait. Lodd. ex Steud.*), lingonberry (*Vaccinium vitis-idaea L.*) is less than 1%. Plant aggregates are characterized by significant species diversity, and the total projective cover amounts to 80%. The layered structure begins to be formed with a clearly defined herbaceous layer (moss and shrub layers are not developed everywhere).

Therefore, recovery of tundra vegetation communities within the drilling site limits requires at least 15 years. At the same time, the fraction of poaceous, cyperaceae, and composites in their composition increases as compared to undamaged tundra areas, while the fraction of typical tundra flora families is significantly reduced.

In the middle taiga subzone of Western Siberia (Tyumen Region), the pioneers of reclaiming the soils on drilling sites are meadow horsetail, hare's tail, various sedge species, European water-plantain (*Alisma plantago-aquatica L.*), fireweed, and certain poaceae. Tree species (aspen, birch, pine, willows) are regenerated.

On drilling sites in the south taiga subzone of the European part of Russia (Perm Region), the initial stage of vegetation recovery involves formation of forb-grass associations with a predominance of tussock grass, smooth meadow grass, plantains, yarrow (*Achillea millefolium L.*), and common dandelion. The plants are low. In all cases, the vegetative cover of the soil surface does not exceed 60% [GLAZOVSKAYA, 1982].

Areas polluted by mineralized effluent water

Under humid climate conditions, vegetation communities in areas polluted by mineralized effluent water recover fairly quickly.

In the central part of effluent water impact sphere, intensive recovery of the meadow community is observed as early as in one year's time, mainly by hoary plantain, silverweed, orange foxtail, and knotgrass (projective cover of 80%). However, all plants were significantly smaller in size as compared to specimens in unpolluted areas. In the forest, the soil cover recovered to a condition close to the initial one in 4 years after the accident.

10.5. TECHNOLOGICAL APPROACHES TO REMEDIATION OF LAND POLLUTED BY OIL AND PETROLEUM PRODUCTS

Main principles of remediation

Heavy pollution results in suppression of natural soil purification and adaptation mechanisms. Remediation techniques should create the conditions required for normal functioning of natural purification mechanisms and its intensification. *Remediation continues the process of natural purification and aims at accelerating this process using the ecosystem natural reserves: climatic, microbiological, and landscape-geochemical.*

Land remediation is a set of measures aimed at recovering the original functions of disrupted and polluted land, improving environmental conditions, and removing the anthropogenic factors that present a hazard to human health.

In many countries, the responsibility for purification and remediation of land polluted by oil and petroleum products is a legal requirement. All entities that produce and process oil, and transport, store, or use oil and petroleum products are bound to take emergency action to clean and remediate polluted land outside their production facilities.

Scientifically justified measures on polluted land remediation, particularly, the soil cover, are unfortunately not yet universally implemented in practice. Therefore, numerous experimental studies on remediation of soils polluted by oil and petroleum products yield ambivalent results. The same measures taken under different conditions result in different consequences. The choice of ameliorants generally does not take into account the consequences of their application. Today, there are different methods applied for cleaning soils polluted by oil and petroleum products. The measures to deal with oil pollution consequences require a scientific justification, which should be based on the following main principle: *not to cause any more harm to the ecosystem than the harm caused by pollution.*

Any concept for polluted ecosystem recovery must be grounded on this principle. Its essence consists in the maximum mobilization of internal ecosystem resources in order to restore its original functions. Ecosystem natural purification and remediation are a continuous biogeochemical process.

Oil pollution differs from many other types of anthropogenic impact due to the enormous quantity of substance that can be released from contamination sources to the environment. Such "burst" load can destroy an ecosystem. When evaluating the consequences of contamination and selecting the land remediation technology, one must firstly determine whether the ecosystem will return to the stable condition or will be irreversibly degrading. In other words, selection of land remediation technology must be based on soil resilience to oil pollution.

Assessment of natural purification mechanisms provides the basis for justification of environmental regulations on soil pollution by oil and petroleum products, and allows one to evaluate modern technologies of land purification and remediation after oil and petroleum products pollution. Without doubt, one should pay attention to drawing up the remediation plan, which is the main form of managing this process. Cleaning up the emergency spills of oil and petroleum products is an emergency remediation measure. The process of eliminating residual pollution takes much more time. During remediation and environmental monitoring of its progress in different natural zones, one must take into account the natural-climatic features of the territory. Developing and implementing "low impact" remediation technologies and focusing on the quality of recovered land must be one of the high priority environmental goals in each region. Before large-scale implementation, remediation technologies must undergo environmental expert assessment.

The rationale for remediation of contaminated soils

Soil is characterized by a high degree of internal processes independence. Therefore, the task of environmentally oriented remediation is to control the processes that naturally recover its fertility.

The options for controlling remediation processes in polluted soils are connected with a wide range of determinant restrictions: resource, biological, ecological (imposing bans on any remedial actions) and, finally, economic.

Resource restrictions are restrictions of "the first order." They include the soil natural properties that primarily depend on the climate. It is quite evident that resource restrictions of remediation processes in the steppe zone and in piedmont and mountainous areas will be different. In dry steppe areas, the restrictions could consist in the deficit organic substance in the soils, amount of precipitation, soils salinization and alkalinity, as well as the risk of ascending migration of salts (secondary salinization) caused by artificial moistening. In piedmont and mountainous areas, it could be the potential moisture excess that facilitates the spreading of oil pollution, while at the same time decreasing the specific pollution. Soil texture and the depth of soil profile could be among the restrictions. The properties of oil-contaminated soil should be taken into account in development of its purification systems.

Biological restrictions make it possible to take into account top limits of the biological processes intensity in the soil. Each type of soil is characterized by its "specific" microbial associations, and the maximum possible level of biochemical activity, which, to a large extent, determines the options for applying bioremediation methods. Biological restrictions are revealed by the minimum time for decomposition of hydrocarbon pollution.

Environmental restrictions include information on soils resilience to pollution. The choice of remediation technology depends on the resources for natural soil purification. During development of bioremediation technologies, all the items that determine the environmental consequences of used technologies are taken into account.

Economic restrictions imply obtaining the highest possible environmental benefit with the minimum production and financial costs. When choosing the technology for soil purification from organic contaminants, one should keep in mind the risk of soil resources loss or deterioration. For example, if the developed technology, on the one hand, is efficient in removing organic pollutants from the soils, but, on the other hand, can result in erosion, salinization or groundwater pollution, reduction of humus horizon depth, etc., then the positive environmental effect of remediation will be significantly canceled out by the deterioration in the landscape condition. If, on the contrary, at the time of soil pollution the source of pollution is eliminated, then the pollutant concentration in soils, for example, petroleum products, will be gradually decreasing and will ultimately reach the safe level. In this case, it is not ecologically and economically feasible to perform special soil remediation work because of the risk to cause even greater damage to the soil ecosystem.

The set of remediation methods for soils polluted by oil and petroleum products will vary depending on the nature of contamination. There are two types of soils and ground pollution:

- pollution resulting from emergency spills of oil and petroleum products during well drilling and operation, pipeline ruptures, tank damage at oil depots or tank-car damage on railways, and in other situations. Oil emergency spill occurs suddenly, exposing the ecosystem to a "shock" impact;
- residual pollution after emergency spill cleanup or pollution from chronic leaks out of faulty equipment or small-scale discharges of oil and petroleum products.

Emergency massive oil and petroleum products spills on the surface result in an extreme pollution level. Spill consequences must be mitigated within the shortest time possible (days, weeks), before the pollutant is completely absorbed in the soil. Pollution that remains after oil or petroleum product gathering is referred to as residual pollution. It can be moderate or heavy.

Moderate pollution can be mitigated within the next 5 years by means of natural purification processes. Heavy pollution takes more time to be mitigated. Special measures are required to speed up the recovery process.

Soils are regarded as subject to remediation if the concentration of polluting substances reaches a level whereby suppression or degradation of the vegetation cover begins, the environmental balance in the soil biocenosis is disrupted, microorganism activity is inhibited (species of algal flora, mesofauna, etc. disappear). Oil products are washed out of soils into ground or surface water, hydrophysical properties and structure of soils change, and the performance of agricultural land is reduced.

Technologies to deal with the consequences of emergency oil spills

The main objective in an emergency situation is to ensure that the spilled oil is localized and reliably isolated in order to prevent hydrocarbon filtration into potable water bodies and into groundwater [KHAUSTOV, REDINA, 2006].

The main urgent technological measures are implemented in compliance with the available plan for mitigation of the consequences of emergency oil spills. The removal of oil and petroleum products from the soil surface and from the plow horizon is ensured by technical facilities and application of adsorbents, absorbents (ameliorants), and chemical agents (high reactivity mixtures). A variety of materials is used as oil sorbents:

- natural minerals: zeolites, dispersed silica, lime, sodium sulfate, iron oxide;
- artificial minerals: perlite, expanded clay, silicagel;
- carbonaceous substances: coal, graphite, peat, shales, sapropel, organic fertilizers;
- natural materials: husks, hay, straw, moss, leaves, sawdust;
- polymers: polypropylene, polyurethane, teflon.

Cleaning efficiency depends on the agent and pollutant reactivity.

Harmful and unacceptable methods for mitigation of the consequences of oil spills

Often, to reduce the time required to mitigate the consequences of a spill, the practice is to mechanically remove topsoil, cover the contaminated surface with sand or earth, or burn oil on the soil surface. These methods are extremely hazardous, as they cause more harm to the soil and landscape on the whole than the emergency oil spill itself.

Removing the contaminated fertile soil layer is identical to its complete destruction, whereas remediation techniques could help save it.

Covering contaminated areas with earth or sand results in pollutant preservation for decades, resulting in an acutely reductive situation on the covered area. Under these conditions, the processes of hydrocarbon compound degradation proceed at extremely low rates, and the processes of light fractions evaporation are also decelerated. Oil that has been preserved in this way becomes a long-term source of ground and surface water secondary pollution. It is also unacceptable to backfill collecting pits and drain trenches where oil has not been completely pumped out. Dry pits are backfilled after weathering of oil residues within 1−2 weeks.

In situ burning of spilled oil or petroleum products should be also attributed to prohibited techniques of oil removal from the soil surface.

This is an inefficient and harmful method. Its application ensures oil disposal only in the surface layer of the soil. At the same time, in the burning areas natural biocenoses are destroyed and atmospheric air is polluted by toxic combustion products. Usually, soil productivity never recovers in the areas of petroleum products burning.

In situ mitigation of residual pollution of soil

Residual pollution can be mitigated in two ways: by removing the polluted soil and sediments with their treatment at purpose-built facilities (ex situ); or by treating soils and sediments directly at the site of pollution (in situ).

When the area of oil-contaminated surface is large, application of ground removal technologies is technically impossible and economically unfeasible. First of all, this results in complete elimination of the soil ecosystem with unpredictable negative consequences for the surrounding territories. Therefore, it is practical to apply remediation technologies directly at the site of pollution. Some of these technologies are listed below [Remediation technology..., 2003].

- integrated agrochemical treatment of the soil to activate the soil microbial cenosis (loosening, plowing, aeration, moistening, application of fertilizers, sowing annual and perennial grasses, introduction of vermiculture). Agroreclamation technologies are aimed at using mechanisms of polluted ecosystem natural purification, therefore they are among the most environmentally friendly and often prove to be an efficient method of remediating the land polluted by oil and petroleum products;
- phytomelioration: sowing plants that are tolerant to oil pollution to enhance the activity of soil microflora;
- pumping out gases and volatile organic substances through vertical and horizontal wells;
- introducing biological, carbon, humic, organic-mineral sorbents into the soil to a depth of between 20–40 cm to 1–3 meters from the surface;
- chemical melioration: improving soil properties by means of introducing chemical substances (chemical ameliorants), applying processes for neutralization, oxidation, and reduction of polluted soils. Lime application, plastering, acidification, and introduction of organic fertilizers are among the most common and efficient methods of chemical melioration. High purification efficiency of ameliorants, for example, applying zeolites, peat, organic fertilizers, etc. is ensured by their ability to simultaneously facilitate the

sorption of oil hydrocarbons and cell adhesion of oil-oxidizing microorganisms in the soil;

- soil purification by applying biopharmaceuticals that contain an association of specific, including indigenous, bacterial cultures, and intensifying their living activity.

Mainly aerobic bacteria are used for the biopurification process. Their ability to actively decompose hydrocarbons has been established, and the requirements to their most favorable growth conditions have been identified. Products of petroleum products microbiological treatment are not hazardous to humans and the environment. Among the hydrocarbon-oxidizing microorganisms, that are most often used in biopharmaceuticals, are bacteria of the genera *Pseudomonas, Rhodococcus, Bacillus, Arthrobacter, Acinetobacter, Azotobacter, Alkaligenes, Mycobacterium*; yeast of the *Candida genus*; actinomycetes of the *Streptomyces genus*; fungi of the genera *Aspergillus* and *Penicillium*, and other micromycetes. Decomposer microorganism destroying oil pollutants are always found in the oil-contaminated soil.

In situ biostimulation always facilitates growth of indigenous microorganisms contained in polluted soil and potentially capable of pollutant disposal. Laboratory tests with polluted soil samples help determine the measures that need to be taken in order to stimulate the growth of microorganisms capable of decomposing this pollutant.

One can determine that the process of oil-contaminated soil bioremediation is completed by the results such as succession of plants and other biota components and normal development of higher plants within one or two growing seasons.

Total length of the remediation process depends on soil-climatic conditions and the nature of contamination. This process takes the least amount of time in steppe, forest steppe, and subtropical regions. It will require more time in northern regions. In different natural zones, the entire remediation process takes up approximately 2 to 5 years or more.

CHAPTER 11

NATURAL PURIFICATION
OF SURFACE AND GROUNDWATER

11.1. SPECIFIC FEATURES OF NATURAL PURIFICATION
IN SURFACE WATER BODIES

Surface and ground waters serve as transit media
for pollutants and are capable of natural purification

Natural purification of surface waters (rivers, lakes, and other water bodies) is the recovery of their natural properties that is completed naturally as a result of interconnected physico-chemical, biochemical, and other processes (turbulent diffusion, sorption, adsorption, etc.).

Self-purification capacity of rivers is closely related to many natural factors: physico-geographical conditions, hydrological regime, activity of microorganisms in water, influence of aquatic vegetation, and others. Table 11.1 lists the types and main processes of natural purification in water bodies.

Table 11.1

**Types and processes of natural purification in water bodies
from polluting substances** [FRUMIN, 2000, P. 81]

Types of natural purification	*Processes*
Physical natural purification	Dilution (mixing), export. Evaporation, sorption by suspended solids and bottom sediments. Sedimentation
Chemical natural purification	Hydrolysis. Photolysis, radical oxidation, redox catalysis
Biological natural purification	Metabolism. Bioconcentration. Biodegradation

Physical processes of natural purification in water bodies

The pattern of main physical processes of natural purification is different in water flows (river systems of various levels) and water bodies (water reservoirs, lakes, ponds).

Physical processes of natural purification of river systems. Main physical factors of natural purification of river systems are dilution, physical dispersion, and export of pollutants outside the water basin limits. The conditions of pollutants dilution are determined by the annual average water flows that allow forecasting the intensity of contamination dispersion in a first approximation. The analysis of cascade landscape-geochemical systems (CLGS), whose methodology was developed by M.A. Glazovskaya [GLAZOVSKAYA, 1988, 2007; GLAZOVSKAYA ET AL., 1983], is very relevant for predicting natural purification of rivers. CLGS incorporates river basins from small water flows to major waterways.

CLGS elements are catchment areas of different level river systems or landscape-geochemical arenas (LGA). LGAs are characterized by intensive atmogeochemical, hydrogeochemical, and biogeochemical processes, as well as by the types of landscape-geochemical geochemical barriers. The nature of the LGAs and CLGS in general provide the basis for predicting the behavior of the anthropogenic substances that circulate in the basins, sites of their accumulation, dispersion, transformation, and effect on environmental components.

The type of CLGS determines the behavior of the anthropogenic substances that enter inside its boundaries: accumulation in final runoff regions, secondary pollution of deltas and estuaries, dispersion across large land and ocean spaces. Polluted water flows in the upper cascade parts can carry nondegraded substances into lower cascade parts. Alternatively, the clean waters of upper cascade parts can dilute the polluted waters of lower cascade parts. The intensity of natural purification of cascade landscape-geochemical systems depends mainly on current speeds and the value of annual runoff from the river basin territory.

The highest water flow speeds that facilitate the enhancement of aeration processes and reduction of organic substance content are typical of mountain rivers. In piedmont areas, the flow intensity may vary depending on regional features and slope characteristics, which, in its turn, will influence the mechanical dispersion of polluting substances.

The correlation between the natural purification rate and the speed of water flow current is a hyperbolic function, which begins to sharply increase from 0.7 m^3/s. In general terms, natural purification of running water depends on the temperature, flow speed, and water mineralization level;

$$K_p = \frac{\sum T > 10\,°C \cdot W}{K_{min}},$$

where K_p is the factor of running water natural purification; $\sum T > 10\,°C$ is the sum of active daily temperatures above $10\,°C$; W is the flow speed; and K_{min} is the mineralization factor [SALMANOV, ISMAILOV, 2014].

The main processes that facilitate the physical natural purification of water bodies are the time of the water body water exchange and the suspended solids sedimentation rate. The sum of annual air temperatures with the stable temperature above 10 °C is of great importance, as it characterizes the possibility of light fractions evaporation from oil films.

The rate of organic substances decomposition increases with growing temperature, and acceleration of catabolites outflow and anabolites inflow. Therefore, natural purification in running water bodies with intensive turbulent mixing of water is faster than in standing water bodies.

Temperature conditions have a positive effect on microflora development and growth and development of aquatic plant communities. The territory climatic potential can be used as a climatic indicator when assessing the ability of water for natural purification.

The sedimentation factor (settling of anthropogenic particles on the bottom with suspended solids) plays a crucial role in the physical natural purification of water bodies from oil and petroleum products. Heavy carbonaceous substances, including oxidized oil components, polycyclic aromatic hydrocarbons, and heavy metals present in the oil composition, are the first to settle. The rate of suspended particles sedimentation and the rate of water mass purification depend on particles size and composition. Anthropogenic substances undergo physico-chemical and biological degradation processes in the bottom sediments. At the same time, sediment detachment can result in secondary pollution of near-bottom waters. High concentration of toxic anthropogenic substances in suspended solids has a negative effect on benthic organisms at the time of settling.

The natural ability of water for purification can be suppressed in two cases. First, when the rate and volume of harmful substances released into water bodies exceed the rate of natural purification. Second, when the harmful substances are inaccessible to natural decomposition processes. This is primarily true for synthetic substances. The processes of natural purification in water bodies (rivers, lakes) from pollution can occur only if the volume of pollution remains below 1/35–45 (2–3%) of the water body volume [ISMAILOV, 2009].

Chemical processes of natural purification in water bodies

Oxygen is one of the main factors that facilitates the process of natural purification of water objects. Water is saturated by oxygen primarily by means of its inflow from the atmosphere and its release by photosynthetic organisms. Oxygen content in water is heavily influenced by wind-driven waves and currents. With the temperature increasing, the content of O_2 in water goes down. That is why rivers are more saturated with oxygen upstream than downstream creating more favorable conditions for their natural purification.

The amount of organic matter available for oxidation in the course of water mass natural purification is evaluated by the biochemical oxygen demand (BOD) and dichromate oxidizability (chemical oxygen demand, COD) ratio. When the process of natural purification is suppressed, a low BOD level relative to the COD is observed [FRUMIN, 2000].

The level of mineralization also affects the processes of natural purification of rivers: increase in total mineralization results in reduced intensity of natural purification. Water mineralization depends on weather conditions. In periods of dry weather, evaporation processes result in salt concentration, whereas in periods of wet weather waters are diluted by precipitation.

In the low streamflow period, the fraction of highly mineralized underground water in rivers inflow sharply increases (up to 40–45%), resulting in a substantially higher mineralization of river water and reduced intensity of their natural purification.

Biological natural purification of water bodies

Biological natural purification of water bodies is ensured by the living activity of plants, animals, fungi, and bacteria and is closely linked to the physico-chemical processes. Biotic processes of natural purification of water include organic substances oxidation and water filtration by hydrobionts [OSTROUMOV, 2004,]. Microorganisms play a special role in oxidizing organic matter, although all groups of aquatic organisms are involved in natural purification of water ecosystems. The habitat of aquatic microorganisms is the water mass and the bottom sediments that are stratified according to physico-chemical properties. They contain different quantities of nutritional elements and provide the medium for development of various microorganism groups — bacteria, protists, aquatic fungi and algae, as well as the water flora. Degradation of oil and petroleum products in water mainly occurs due to the living activity of oil-oxidizing and saprophytic bacteria. The process of oil degradation begins immediately after it is released into a water body. The number of microorganisms increases dramatically reaching its maximum on the 3rd-4th day. Microbiological processes result in destruction of oil film and the oil in water mass, decrease of oxygen concentration in water, and increase of carbon dioxide content. As the quantity of oil is reduced, the number of bacteria gradually decreases.

Biofilms are formed on the surface of solid particles below 0.1 mm in size (silt–sand–fine gravels–coarse gravels–boulders). The smaller the particle, the larger the area of the biofilm formed on its surface and the higher the intensity of water natural purification and the smaller the distance required for natural purification of a river from an organic pollutant.

Besides microorganisms, aquatic vegetation also plays an important role in natural purification of water bodies. This is due to a combination of factors: it releases oxygen in the course of photosynthesis; it is capable of intensively using biogenic elements (first of all, phosphorus and nitrogen); the surface of plants submerged in water is populated by a complex of various organisms participating in degradation of the pollutants: bacteria, fungi, epiphytic algae, protozoa, nematodes, sponges, insect larvae, mollusks, etc.

With respect to organic pollutants, semi-aquatic plants perform the following functions:

1. Mechanically retain suspended and poorly soluble organic substances.

2. Mineralization and oxidation.

3. Detoxication.

The role of aquatic plants in the processes of natural purification of water consists in following:

• stimulate the activity of the microorganisms that dwell on their surface and in the water by metabolism products;
• create an active adsorbing surface;
• maintain a high oxidation level of the environments by enriching water with dissolved oxygen.

Higher aquatic plants accelerate bacterial decomposition of oil and petroleum products by 3–5 times. Various types of oil (crude, salable, emulsified), as well as petroleum products, at a concentration of 1 g/l, take 5–10 days to disappear in the presence of plants, while it takes 28–32 days without plants [SADCHIKOV, KUDRYASHOV, 2004].

11.2. PROTECTION AND PURIFICATION OF GROUNDWATER

Groundwater is protected from pollution if:
• this water is isolated from surface infiltration water by thick water-resisting or low-permeability rocks;
• infiltration water is being completely purified from oil contaminants with the participation of biogenic and abiogenic factors in the course of filtration through the mass of rocks and sediments.

The physical, chemical, and microbiological processes occurring in the zone of aeration are able to mitigate the negative effect of infiltration on the subsurface hydrosphere, although to a limited degree. However, some part of contaminants ends up in groundwater, especially if it is located near the surface. Oxidized derivatives of hydrocarbons (acids, alcohols, aldehydes) migrate the most aggressively, as they are easily soluble in water, and reach the groundwater horizon.

Groundwater classification in terms of the degree of protection is based on indicators such as the thickness and lithological composition of low-permeability sediments (determine the rate of water filtration), and groundwater depth level [GOLDBERG, 1998]. If the thickness of low-permeability sediments in the zone of aeration is < 5 m, sandy-loam, and lightly loamy sediments with the filtration factor of > 0.01 m/day are predominant and the time of pollutant filtration to the level of groundwater is 1–10 days, groundwater is categorized as poorly protected. On the Apsheron Peninsula, groundwater in the majority of petroleum-producing areas falls under the poorly protected category [ISRAFILOV, LISTENGARTEN, 1978].

The territory climatic characteristics (annual precipitation amount, their seasonal fluctuations, evaporation level) are among the important factors that determine the possibility of pollutant dispersion in the subsoil mass, which affects the migration ability of hydrocarbons in the ground and the degree of natural protection of groundwater from contamination.

The natural processes that result in purification of polluted surface and groundwater differ in terms of behavior and intensity. Surface water and soil are in contact with the atmosphere and contain microorganisms. Organic pollutants, subjected to dissolved oxygen, light, and elevated temperatures are intensively oxidized, decomposed, and mineralized. Groundwater self-purification is of a different kind.

Groundwater typically has a much lower temperature than surface water. It contains a relatively small reserve of dissolved oxygen and a limited amount of microorganisms, which is why microbiological processes here are less intensive than in surface water and soils. In subsoil horizons, natural purification from pollutant is ensured by oil sorption in rocks, ion exchange, and, less often, by oxidation and decomposition of polluting substance. Porous and clayey rocks absorb oil and its components easily.

Intensive water exchange with surface water results in oil and petroleum products dilution, transfer, and dispersion, as well as partial decomposition, in the course of oxidation-reduction reactions.

Groundwater pollution causes irreversible damage to the biosphere, depleting the reserves of natural potable water. Groundwater almost never completely purifies by the natural way, as the oil components absorbed by mineral particles of water-bearing layers turn into a long-term source of water flow secondary pollution.

Purification of polluted subsurface water is a costly measure and is not always possible. Therefore, all necessary actions must be taken to prevent its pollution. Purification of subsurface water can achieve its goal only after the pollution source contained in the soil is eliminated [HARRIS, 1990].

CHAPTER 12
RESILIENCE AND PURIFICATION
OF THE MARINE ENVIRONMENT

12.1. NATURAL PURIFICATION
OF THE MARINE ENVIRONMENT FROM
OIL AND PETROLEUM PRODUCTS

Abiogenic factors of natural purification
of the marine environment

Marine ecosystems have a significant tolerance toward external impacts, which enables them to withstand the formidable global anthropogenic load that has been created and is intensified by economic activity of humans.

The resilience of marine environment petroleum anthropization is ensured by a variety of natural mechanisms. They help mitigate the negative impact of anthropogenic hydrocarbon flows on marine ecosystems and facilitate subsequent natural purification of the environment from pollutants.

Hydrophysical, sedimentational, geochemical, and biogeochemical natural factors play a dominant role in natural purification of the marine environment from oil pollution.

Hydrophysical factors—structure and seasonal dynamics of waters, changes in physical properties of water in dependence on the solar radiation and atmospheric conditions, runoff amount, and other geophysical phenomena—create a powerful mechanism of dispersion of pollutants reducing their concentrations until the geochemical anomalies in the aquatic environment are completely eliminated.

The studies of hydrophysical factors of natural purification conducted by Russian researchers in the Caucasian part of the Black Sea revealed the importance of the seasonal dynamics of seawaters for natural purification of the sea. In the summer period, due to high temperature of the surface layer, weak wind-driven circulation, and weak currents, vertical exchange between surface and deep water in the Black Sea is less intense. Anthropogenic pollutants are concentrated in the upper 10- to 15-m-thick water layer. Pollutant dispersion and natural purification are decelerated at this time. The environmental situation is particularly stressed in the coastline zone where the pollutant concentration is the highest. In the winter period, the invasion of cold air, cooling of water top layer, and stronger winds result in intensive mixing of water. In the newly formed homogeneous convection layer having a thickness of 30–130 m, anthropogenic pollutants are actively diluted and dispersed, which ensures their efficient biogeochemical degradation. The inflow of clean deep water into the upper water layers plays an essential role. The surface layer is renewed and cleaned of polluting substances.

The effect of sedimentation factors consists in involvement of poorly soluble pollutants into the process of sedimentation. These pollutants include heavy components of oil, petroleum products, and the majority of PAHs. However, although water is cleaned when pollutants are settled on the bottom, they create persistent anthropogenic geochemical anomalies in bottom sediments. The formation of geochemical anomalies is particularly intense against the background of increased sedimentation rates in shallow sea areas, for example, in closed and semi-closed bays. Further behavior of these anomalies depends on the biogeochemical processes and the possibility for mixing of the upper layer of bottom sediments (Fig. 12.1).

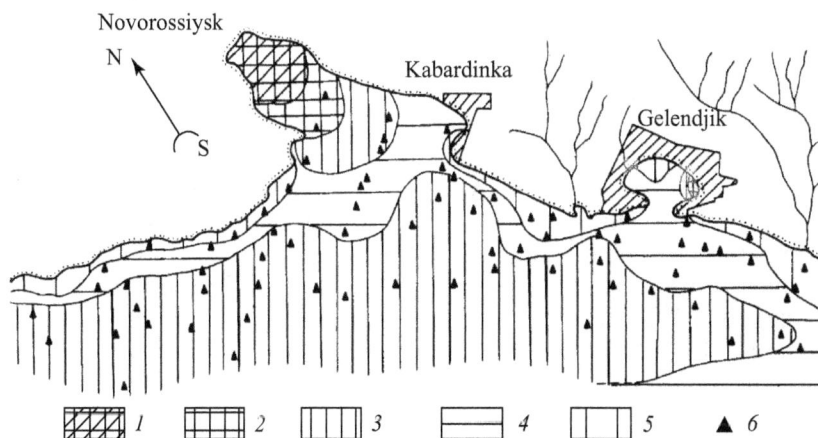

Fig. 12.1. Distribution of total petroleum products in the surface layer of bottom sediments in the offshore area of the Black Sea between Novorossiysk and Gelendjik [TECHNOGENIC POLLUTION, 1996, P. 323]. Content of n-hexane-soluble petroleum products (mg/kg): *1* — more than 1,000; *2* — 200–1,000; *3* — 50–200; *4* — 30–50; *5* — less than 30; *6* — sampling points.

Geochemical factors of self-purification of the marine environment include physico-chemical processes of hydrocarbon oxidation on the water surface and their settling on natural or human-made geochemical barriers. Oxidation efficiency depends on the quantity and duration of solar radiation. As calculated by K.A. Parker and co-authors [PARKER ET AL., 1971], 2 tons of oil per 1 km^2 in the surface layer of water can be oxidized by sunlight in 24 hours. Natural geochemical barriers in seawater are mineral suspended solids that sorb the high molecular weight components of oil and petroleum products. This results in formation of oil aggregates that sink to the bottom in the form of bitumen lumps, thereby cleaning the water mass.

Biological factors of natural purification of the marine environment

Natural purification of the marine environment from oil and petroleum products is ensured collectively by hydrobionts of all trophic levels, particularly, by hydrocarbon-oxidizing microorganisms (HOMs).

The data available on the distribution and species diversity of hydrocarbon-oxidizing microorganisms help assess a water area capacity for natural purification from oil. According to calculations, the capacity for oil degradation along the domestic coast of the Black Sea within the 100-meter isobathic line is estimated at 2,000,000 tons per year. According to an American researcher R. Atlas [ATLAS, 1993], the number of bacteria and fungi using oil as a nutrient medium, i.e. the number of hydrocarbon-oxidizing microorganisms, in clean seawater is not more than 0.1–1.0% of the heterotrophic bacterial communities population. In oil-contaminated water, the number of HOMs can increase up to 1–10%. The rate of biodegradation depends on several factors, primarily, on oil and petroleum products composition (simple low molecular weight compounds degrade faster than high molecular weight compounds), degree of their dispersion, ambient temperature, species composition, and number of oil-oxidizing microorganisms.

Protozoa develop in oil released into the sea along with bacteria. According to O.G. MIRONOV [1992], the activity of infusoria contributes to more complete transformation of emulsified oil in seawater. A drop of oil is attacked by hydrocarbon-oxidizing bacteria on the oil-seawater interface. In the course of bacteria development, the bacterial film is formed that surrounds the oil drop and consists of living cells and bacterial detritus. Infusoria feed on detritus and thereby ensure a more efficient penetration of microorganisms into oil. Incidents of crude oil ingestion by infusoria have also been observed.

Zooplankton behaves in a similar manner, ingesting micro-fine oil particles in the process of nutrition. Oil emulsion drops have been found in bodies of organisms studied under a fluorescent microscope. Aromatic hydrocarbons were identified in the oil composition with their content amounting to 175 µg/kg of dry substance. At oil concentration in water of 10 mg/l, its content in fecal matter was almost 50%. Zooplankton samples contained substituted naphthalenes.

Filtration activity of certain zooplankton species, for example, Calanus finmarchicus Gunn copepod, can reach 15 l/day per one specimen, i.e. each organism can ingest up to $1.5 \cdot 10^4$ gm. of oil daily. Calculations demonstrate that a population of 2,000 specimen/m^3 in density dwelling on the surface of 1 km^2 in area and 10 meters in depth can remove 3 tons of oil per day from the water body [Mironov, 1992].

Among other organisms that play an important role in purification of seawater from oil pollution are filter-feeding bivalves — mussels. Oil is accumulated in mussel organisms. They can filter emulsified and suspended hydrocarbons from seawater. These hydrocarbons pass into the bound state in the composition of feces and pseudofeces. These metabolism products of mussels positioned in seawater with oil contain a high number of hydrocarbons (315.5 and 242 mg / 100 gm of dry substance, correspondingly). After mussels have been placed into clean seawater, hydrocarbons are gradually removed from these hydrobionts.

Filter-feeding bivalves not only mechanically remove oil from seawater, but also transform it. Oil residues extracted from mussels' excretory waste contain more heavy oil fractions and more aromatic structures within the lube fraction compared to the original oil. A shift towards heavier alkanes and isoprenoids is observed in the aliphatic hydrocarbons composition.

Hydrobionts involved in the system of seawater and bottom sediment biological treatment also include various polychaetes that are able to metabolize oil hydrocarbons, including even polycyclic aromatic hydrocarbons that are resistant to degradation.

The ability of filter-feeding hydrobionts to purify seawater should be used for accelerating purification of contaminated seawater areas by creating and expanding their natural habitats.

Mussels growing experiments revealed that in the vicinity of Sevastopol 65% of mussels located on a collector (nylon strap) grew to 25—35 mm long and more within the period of 9 months, between March and December. Their filtration capacity is approximately 17 l/day. This helps calculate the number of organisms required to filter a given volume of seawater. Tests revealed that when oil-containing seawater was treated using Mediterranean mussels, the concentration of diesel fuel proved to be twice as low as after the treatment in a unit without mussels [Mironov, 1992]. Furthermore, the largest quantity of hydrocarbons is found in pseudofeces independent of the initial toxic substance concentration (1.0–25.0 mg/l).

12.2. TECHNOLOGICAL APPROACHES TO THE CLEANUP OF THE MARINE ENVIRONMENT AFTER OIL SPILLS

An oil spill is one of the most environmentally hazardous consequences of marine anthropization. All existing models for environmental risk and environmental damage assessment are primarily based on analyzing oil spill consequences. Most often, oil spills at sea occur when it is transported by vessels or pumped through pipelines. However, large accidents can also happen in the course of field surveying and operation. Such an accident occurred in 2010 when drilling a deep well in the Gulf of Mexico. The flowing of the oil well located on the 1.5 km deep seabed could not be contained for several months.

The goal of oil spill prevention at sea is to minimize the damage caused to natural resources and to social and economic activities, as well as to reduce the time required for recovering the damaged resources by implementing acceptable cleanliness standards.

In practice, the choice of technical solutions for spill localization and mitigation is rarely confined to just one method. Typically, a variety of technologies is applied in different impact zones, including mechanical, physico-chemical, and biological. The list of recommended methods includes manual cleanup, using dispersants and sorbents, biological purification, mechanical regeneration, and burning.

Mechanical methods mainly consist in setting-up boom containments and mechanically collecting oil inside contained areas. In shallow waters, oil is usually removed using oil skimmers.

Successful application of mechanical oil collection methods in offshore areas is normally prevented by high waves and sea currents, whereas on shallow marine shelf the situation is aggravated by the vicinity of poorly protected ecosystems in the tidal zone and coastline. This limits the time available for spill mitigation activities dramatically.

In exceptional situations, when there is a risk of oil quickly spreading on the shore or if a special permit is received from the authorities, oil spills seat sea and on shore can be mitigated using the method of in situ burning of oil on water. It is applied only if the thickness of oil film is at least 3 mm. Burning of residual heavy fractions is prohibited because the thick black smoke that is released contains large quantities of PAHs and heavy metals, nitrogen oxides, and sulfur.

Physico-chemical methods are applied when it is impossible or inefficient to collect oil mechanically. Applying dispersants has become the most popular among these methods. Dispersant application allows to remove oil from the water surface, reduce migration ability of oil stains and tar balls, and treat extensive water areas in a short time using aviation. This decreases the risk of damage to birds and mammals and disintegrates oil accelerating its biochemical decomposition. Although there are a few contra-indications for dispersant use (meaning that they must be applied with special care), sometimes this is the only method that is capable of preventing an environmental disaster. Applying dispersants in deep water areas is considered as the most environmentally safe option. Within the limits of shallow shelves, artificial oil dispersion can cause severe damage to hydrobionts.

Oil biological decomposition can proceed fast enough only if the oil amount is not large. Otherwise, the oil film prevents access to oxygen, and development of microorganisms decelerates.

Shallow shelves and offshore landscapes are the most vulnerable areas of marine ecosystems. At the same time, these areas feature the highest level of biological productivity and biological diversity. The time for which oil pollution is maintained in coastal landscapes depends on their resilience capacity. A rocky cape can be cleaned from oil within a few months. On a sandy and cobbly beach or on an open low tidal shore, the pollution can hold for 2 years; on a rocky shore and closed low tidal shore — up to 5 years; while on sea marshes, it can hold for up to 10 years [REVELL, REVELL, 1995].

Technological approaches to methods of seacoasts purification from oil pollution include oil collection using technical equipment, sorbents, and pharmaceuticals stimulating the development of hydrocarbon-oxidizing bacteria. Applied practice includes removing oil-contaminated sediments and vegetation, washing oil off by water at high pressure and/or high temperature, using sandblast cleaning systems, chemical purification if there is a risk of unique ecosystems being completely destroyed or there is serious danger for human health. In particularly vulnerable landscapes, for example, within the limits of salty marshes, the options of landscape natural purification have priority. Interference is required only in emergency situations when oil is forecast to spread across extensive areas or there is a pollution hazard for birds and other animals.

In any case, methods of coast protection and purification after oil spills at sea must be chosen based on the same principle that is applied for soil protection: not to cause more harm to the ecosystem than the harm caused by oil pollution.

PART FOUR
DIAGNOSIS, MONITORING, AND FORECASTING OF THE PETROLEUM ANTHROPIZATION EFFECT ON THE ENVIRONMENT

CHAPTER 13
CHEMICAL AND BIOLOGICAL DIAGNOSIS OF ENVIRONMENTAL POLLUTION

13.1. INSTRUMENTAL CHEMICAL-ANALYTICAL DIAGNOSIS OF ENVIRONMENTAL POLLUTION

General description of methods

Although the problem of oil and petroleum products diagnosis in soils, rocks, and water exists for a long time, it is still far from being solved definitively. This is explained by the complex composition and high diversity of the substances that are incorporated into the term "oil and petroleum products", as well as by the ambivalence of the natural environment responses to the effect of these substances. However, proper quantitative and qualitative diagnosis of environmental components pollution by oil and petroleum products is of crucial importance for the choice of environmental protection measures.

Substances that are referred to as petroleum products by environmental scientists circulate in the environment and are diverse. It is very difficult to give them a common analytical definition. Oil of various compositions, gasoline, kerosene, diesel fuel, various lube oils, residual fuel oil, road bitumen, paraffin wax have no other common property than complete or partial solubility in organic solvents. Moreover, even optimum solvents will be different for different types of petroleum products.

In many countries, the total petroleum hydrocarbons (TPH) are used as a universal means for petroleum products diagnosis. This term describes the totality of saturated, unsaturated, and aromatic hydrocarbons that are subjected to further detailed studies. TPH is obtained by removing high molecular weight and bioorganic components from extraction products using different solvents and columns with mineral sorbent.

This approach to studying "petroleum products" in soils omits from consideration the high molecular weight resinous-asphaltenic substances, heavy polynuclear polycyclic hydrocarbons (studied separately), and the bulk of toxic chemical associate elements that have a negative impact on soils and ecosystems in general. At the same time, these particular components are most often bound in upper soil horizons, dramatically modify their hydrophysical properties, and release toxic substances into the soil environment. Their negative effect is much stronger than the effect of light hydrocarbons, particularly if they penetrate deep into the soil profile.

One of the critical tasks for the diagnosis of petroleum and associated chemical pollution of soils is to include resinous-asphaltenic and heteroatomic oil components into the analytical concept of "oil and petroleum products", and to develop environmental regulations for safe level of soil pollution, as well as unified methods to determine specific thresholds for each type of chemical pollution.

A variety of physico-chemical methods is used to study petroleum products in soils by environmental laboratories of different entities [PIK-OVSKII AT AL., 2017). Each method has its own field of application and solves a specific range of tasks. Understanding the behavior, origin, and environmental significance of hydrocarbons in soils is much improved by studying the whole complex of gaseous, liquid, and solid hydrocarbons that occur, including high molecular weight substances chemically linked to them that are united under the concept of "hydrocarbon state of the soils."

The procedure for hydrocarbon state of the soils research includes studying and interpreting the following forms of carbonaceous compounds in soils:

- noncarbonate ("organic") carbon;
- natural gases C_1-C_4 (free and retained);
- bitumoids — total substances extracted by nonpolar and low-polar organic solvents (n-hexane, chloroform);
- total hydrocarbon compounds of various classes (total petroleum hydrocarbons);
- individual compounds present in the bitumoid (alkanes, polycyclic aromatic hydrocarbons, polycyclic cycloalkanes, etc.).

After extraction, the bitumoid is studied as a whole before its separation into individual parts ("oil and grease method"). The composition of bitumoids includes hydrocarbons and a wide range of simple and high molecular weight heteroatomic compounds. The ratio of bitumoid main components (lube oils, resins, and asphaltenes) is the major characteristic of a bitumoid. In Russia, this method has been termed "component analysis" [FLOROVSKAYA, 1975], while in the West it is referred to as the

SARA method (saturates, aromatics, resins, and asphaltenes) [ASKE ET AL., 2001; FAN ET AL., 2002].

In the extensive toolkit of environmental diagnosis (chromato-mass-spectrometry, gas, gas-liquid, liquid chromatography, spectrophotometry, fluorometry, spectrofluorometry, etc.), luminescent methods are the ones that stand out. Despite their limited field of application and complexity of analytical interpretation, they are highly instrumental for conducting comprehensive geochemical and environmental research.

Luminescent methods for an oil and petroleum products analysis are essentially luminous excitation of substance by ultraviolet rays and receiving the emitted light in the ultraviolet and visible spectrum. By studying the emission (fluorescence) intensity and spectral characteristic, one can perform a quantitative and qualitative substance analysis at extremely low concentrations in solutions.

Luminescent methods are primarily applied for diagnosis and quantitative analysis of oil, petroleum products, bitumoids, and polycyclic aromatic hydrocarbons (PAHs) present in natural environments, as well as for studying the effect of these substances on the environmental situation.

Luminescent methods have a few distinctive features.
- the analysis does not require to destroy the object or expose it to thermal impact which could modify its chemical structure;
- the sample is preserved throughout the entire research cycle. The same measurements can be repeated many times, and the same sample can then be analyzed using other methods;
- very high sensitivity provides for visual identification of oil concentration in solution of $1 \cdot 10-6$ g/mL, while individual compounds can be instrumentally recorded at concentrations of $10-10-10-11$ g/mL;
- speed, possibility of mass measurements;
- continuous observations of object luminescence both in space and over time are possible;
- remote measurements are possible;
- methods applied in field conditions;
- as living organisms, along with oil and petroleum products, can have luminescent properties, luminescent analysis can be used for comprehensive ecosystem research.

Luminescence-bitumen analysis was developed by V.N. Florovskaya and V.G. Melkov to study oil residues in rocks with the objective of searching for oil and gas fields [FLOROVSKAYA, 1975]. Subsequently, this method has been adapted for diagnosis of soil pollution by oil and petroleum products. The method implies fluorometric measurement of bitumoid solution luminescence in the ultraviolet and visible spectra, as well as measurement of a luminescent zone color and width on a strip of chromatographic paper (capillary extract). Solution concentrations are calculated by means of selected standard solutions which luminescent properties are similar to the measured substance. Analysis result depends on the correctly chosen reference. If an appropriate standard cannot be found in the database, it is prepared from a larger sample of the same soil according to a special procedure. Method sensitivity is $1 \cdot 10^{-6}$ g/mL.

Identification and quantification of polycyclic aromatic hydrocarbons are ensured using the low temperature luminescent-spectral method (Spolsky spectroscopy) in spectrofluorometers at nitrogen freezing temperature ($-196\,°C$) or lower [ALEKSEYEVA, TEPLITSKAYA, 1981; GENNADIEV, PIKOVSKIY, 1996; ROVINSKY ET AL., 1988]. The method is based on the E.V. Shpolsky's discovery of quasi-line spectra of individual aromatic molecules luminescence in a frozen matrix of n-alkanes. Identification sensitivity of an individual compound in a solution is $10-10-10-11$ g/mL. PAHs identification and quantitative analysis was performed by means of comparison to a certified reference solution of a mixture of different PAHs (for example, international standard 2260a certified by the National Institute of Standards and Technology, USA). If heavily diluted solutions are to be analyzed, PAHs chromatography can be replaced by the "spectral fractionation" method based on the choice of the most suitable spectral length of excitation and fluorescence waves for each compound.

Any method, including luminescent analysis, has its limits when it comes to the range of applications and its features. The point is to find an optimum set of methods that is feasible for solving a specific task.

13.2. BIOLOGICAL DIAGNOSIS
OF ENVIRONMENTAL OIL POLLUTION

Approaches to biological diagnosis of oil pollution

Biological diagnosis methods are essential for assessment of environmental pollution, as they provide a direct indication whether the ecosystem condition has changed due to the impact of anthropization. Two approaches are identified for oil pollution biodiagnosis, i.e. bioindication and biotesting.

Biological indication of the level of environmental pollution by oil and petroleum products consists in studying how living organisms respond to pollution; the study is carried out directly in their habitat. Depending on their tolerance toward oil under natural conditions, the biocenoses and individual specimen respond differently to the same pollution level and type of pollutant. Biological indication detects the reaction of the ecosystem to integral pollution under given conditions. Therefore, biological and geochemical pollution indication must be used as a whole.

Organisms from different systemic groups are used for bioindication of oil-contaminated environments: higher plants (phytoindication), algae (algoindication), microfungi (mycoindication), and animals (zooindication). Specific procedures often apply different systemic groups of organisms as bioindication objects.

Biological testing is conducted in laboratory conditions. It is based on placing a living organism, its tissues, or a combination of organisms into the component of natural environment (water, soil, bottom sediment) which condition is to be tested. Test object response under reference conditions defines the level of hazard that this environmental condition represents for living organisms.

Below, we consider certain methods of environmental oil pollution bioindication and biotesting.

Method of environmental pollution phytoindication

Cenotic level of indication. The objects of bioindication are higher plant communities. Indicator properties of plant communities are identified by comprehensive processing of a large number of geobotanic descriptions, also using the floristic classification method by Braun-Blanquet [BRAUN-BLANQUET, 1964] and scale ordination by L.G. Ramensky [RAMENSKY ET AL., 1956].

Population level of indication. The objects of bioindication are populations of indicator plant species. Species with a high occurrence on the studied territory can be used to indicate oil pollution. Oil pollution indicator species are identified using a direct gradient analysis of species (i.e. analysis of their distribution along the pollutants gradient that includes the calculation of their phytocenotic role at each grade of the pollution level) and a rank correlation analysis (rank correlation of species abundance and pollutant content in the soil). For example, in Kaliningrad region typical meadow plants can serve as an indicator of the soil pollution level by oil. Most weed species can be used to indicate the degree of substrate mechanical disruption in the territory of oil fields. Soil anthropogenic salinization is indicated by halophilic species — weeping alkali grass (*Puccinellia distans*) and black grass (*Juncus gerardii*) [NERONOV, 2008A].

A promising method of phytoindication in oil field regions could be studying the fluctuating asymmetry (deviations from perfect bilateral and radial symmetry) in populations of indicator plant species. Plants with pronounced bilateral symmetry of lamina are used as indicators: goatweed (*Aegopodium podagraria*), common coltsfoot (*Tussilago farfara*), greater plantain (*Plantago major*), bastard clover (*Trifolium hybridum*), and creeping clover (*T.* repens); among agricultural plants, it is barleycorn (*Hordeum sp.*), oats (*Avenna sp.*), and wheat (*Triticum sp.*). Tree plants that are most often used for indication by fluctuating asymmetry method are warty birch, balsam poplar (*Populus balsamifera*), Norway maple (*Acer platanoides*), and Canadian maple, among aquatic plants — pondweed species (the *Potamogeton genus*) [MELEKHOVA O.P., YEGOROVA E.I., 2007].

Method of environmental pollution algoindication

Algae have a few advantages as indication objects over heterotrophic microorganisms: relative ease of cultivation and species identification and environmental behavior of individual species is well studied [SHTINA, 1990].

Algoindiciation of environmental pollution (water, soil, and sediments) by oil is highly efficient on the community level. To assess the condition of algal communities, the indicators generally accepted in algology are used (total species diversity, species composition, ratios between algae of different taxonomic and ecological groups, composition of dominant and characteristic species, species number). Additional options for indicating the direction of physico-chemical processes (alkali-acid and oxidation-reduction conditions, anthropogenic salinization) in oil-contaminated soils and water are provided by an analysis of the diatom algae complex properties for which the largest body of environmental knowledge has been accumulated.

To identify the levels of soil pollution by oil, one must determine the nature of the algae communities' response to different pollutant doses (Chapter 5).

To ensure algoindication of levels of soil pollution by oil, simulation field experiments using the "dose-effect" principle were conducted. The basis of soil pollution levels assessment is the quantitative survey data on algae ecological and morphological groups (i.e. in more detail than is common practice in the majority of algological researches). This approach made it possible to compare the reaction to oil impact in algae communities with various species composition coming from different soils. As a result, information on the critical level of anthropogenic load for soil algae and on the range of oil concentrations in which they can sustainable develop was obtained. It has been established that the algae community response to soil pollution by oil is determined by residual oil content in the root habitable layer that depends not only on the type of soil but also on its morphological traits, terrain position, land use type, and other factors [DOROKHOVA, SOLNTSEVA, 2012].

Algoindication helps tackle many tasks. For example, in surface ecosystems analyzing the parameters of soil algae communities allows to ensure indication of:

- tendency of anthropogenic modification of soil and sediment properties;
- duration of acute toxic effect of oil on soil microbiota (by means of field simulation of soil pollution-natural purification processes using the principle known as the dose-time-effect relationship);
- levels of soil pollution by oil;
- efficiency of remediation methods for oil-contaminated soils.

Algological studies of soils on dated crude oil spills and oil-contaminated effluent water in the Kama region near Perm revealed that each stage of anthropogenic transformation of the initial soddy-podzolic soils (Glossic Retisols)—from anthropogenic bituminous saline land in the first year after contamination to secondary solodized soils (Natric Retisols) in 18–20 years after contamination) correlates with the specific algae communities [YELSHINA, 1986; SHTINA, NEKRASOVA, 1988].

Method of zooindication of environmental pollution

Cenotic level of indication. Pedobiota communities can be used to indicate oil pollution of soils and sediments. Recorded indicators are total number of micro- and mesofauna, species diversity, the animal taxonomic groups ratio within these size groups. Mesofauna representatives, oribatid mites, and springtails are sensitive indicators of oil pollution at the first and second stages of oil degradation in soils. These stages are characterized by their numbers repression, reduction of their fraction in the small invertebrates complex, severe degradation of species composition, change of dominating species composition, appearance of species typical for anthropogenically damaged ecosystems [ARTEMYEVA, 1989].

Large invertebrates (primarily earthworms) are used at the third stage of oil degradation. For more reliable bioindication of oil pollution, it is recommended to use organisms from a different systemic group that belong to different links of the trophic chain.

Population level of indication. The population level involves studying indicators, such as the size of indicator animal species population, their sex-age structure, life expectancy, fluctuating asymmetry of specimens, etc.

Changes in the individual and population characteristics of two indicator invertebrate species — Prussian carp (Carassius gibelio Bloch) and lake frog (Rana ridibunda Pall.)—were used to assess oil pollution levels in the water of the Don River [KARMAZIN, 2010]. Oil concentrations in the range of 0.01−0.5 mg/l (0.2−10 MPC) had a toxic effect on these animals, which was registered by population indicators. For example, at this level of water pollution by oil, the lake frog population suffered from 60−100% frog larva mortality, reduction of their length (at 46th and 57th developmental stage), as well as retarded completion of certain ontogenesis phases. The same reactions to water pollution by oil were observed in the larva of Prussian carp. Moreover, their morphological anomalies were detected.

Organism level of indication. On the organism level, the morpho-anatomical, behavioral and physiological and biochemical characteristics of indicator animal species are studied. Behavioral and physiological and biochemical characteristics of animals' status are particularly sensitive to environmental pollution [Biologic monitoring..., 2007].

The studies in the mouth of the Don River supplemented by experiments [Karmazin, 2010] revealed a high sensitivity to water pollution by oil in the lake frog (Rana ridibunda Pall.) population that is very common in fresh water bodies. The following physiological and biochemical parameters that have demonstrated a clear correlation with the oil dose are proposed as indicators of water pollution by oil: level of oil accumulation in kidneys, number and morphology of blood cells, rate of lipid peroxidation in various organs of this animal.

Physiological−geochemical monitoring of environmental condition

In the 1980−1990s, V.A. Veselovsky and V.S. Vshivtsev from Lomonosov Moscow State University developed a comprehensive physiological−geochemical method for monitoring ecosystem conditions y [VESELOVSKY, VSHIVTSEV, 1988].

This method is based on observing the photosynthetic activity of photoautotroph living cells in soils and aquatic environments with a simultaneous analysis of soil and water pollution by oil and petroleum products. Photosynthetic activity is measured using the phenomenon of long-lived afterglow (LLA) which spontaneously occurs in autotrophs during photosynthesis. The research physiological model used in method development was the Anabaena variabilis Kütz cyanobacteria.

LLA was detected using an installation consisting of a phosphoroscope, excitation light focusing system, and the dark chamber where the object in a temperature-controlled quartz cell was placed.

Figure 13.1 illustrates the experiment schematic, as well as the modeled induction curve and the LLA temperature dependence of cyanobacteria cell slurry developed under normal conditions. At zero time, the light in dark chamber is switched on (arrow 1). LLA is excited to the level ("c") that is proportional to the number of "living" cells. Then, photosynthesis begins. At its induction phase, oxygen is released and LLA is dimmed to the level "a." The value of ($c - a$) was assumed to be proportionate to photosynthesis. At the temperature of 28 °C, the culture begins to be gradually heated (arrow 2), the LLA level is again growing until it reaches the maximum level ("b"). After that, the release of oxygen stops and the photosynthesis decelerates. As the changes in the glow level ($b - a$) / b ratio at the temperature of 28 °C correspond to the changes of value ($c - a$) / c, it has also been used as an indicator of photosynthesis level relative value.

The shape of LLA curves is the same for all the photoautotrophs that develop under their normal conditions. Therefore, the method can be applied for studying the geochemical situation and the reaction of autotroph photosynthesis to its changes.

The physiological–geochemical method was tested on higher plants and aquatic algae in different natural zones (Colchis Lowland, Permian Kama region, offshore area of the Baltic).

The completed research led to several important conclusions.

1. The photosynthetic activity of phototrophic organisms dwelling both in aquatic and terrestrial environments measured by luminescence phenomena is highly sensitive with regard to the quantity and composition of oil released into the ecosystem (Fig. 13.2).

2. Heavy oil fraction has the strongest effect on photosynthesis.

3. The inhibiting action of oil and its individual fractions on the photosynthesis process are more pronounced at lower temperatures.

4. The method under consideration allows assessing the intensity of natural purification of the environment contaminated by oil. For example, the study of Cladophora sp. photosynthetic activity after the Globe Asimi tanker accident confirmed that the functional state of the alga photosynthetic apparatus was nearly normal and that the water area had naturally purified within the elapsed time period.

5. Under the oil pollution conditions, autotroph organisms have the ability of adapting to the new conditions to a certain extent. If their level of photosynthetic activity in polluted environment remains above 40–60% of the baseline, it is highly probable that the plant will leave the stressed state and adapt.

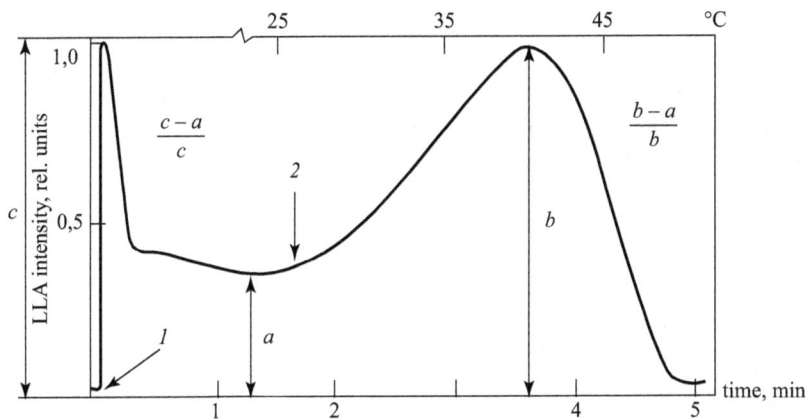

Fig. 13.1. Photosynthetic activity model curve of Anabaena variabilis Kütz cyanobacteria cell slurry. Induction curve and LLA temperature dependence [VESELOVSKY, VSHIVTSEV, 1988, P. 102]. Vertical axis: c — explosive induction amplitude; a — LLA stationary level, b — LLA temperature maximum. Arrows: 1 — start of cell suspension illumination; 2 — start of culture heating.

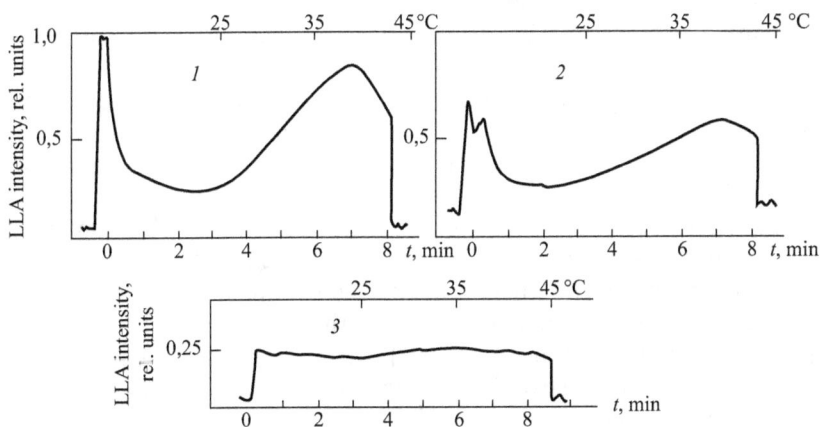

Fig. 13.2. The changes in the photosynthetic activity curve of Hungarian bromegrass seedlings (20–25 days after sowing) depending on the level of soil pollution by oil (a year before sowing). Rates of oil application: (1) 8 L/m² (normal photosynthesis), (2) 16 L/m² (photosynthesis 50% of normal), (3) 24 L/m² (photosynthesis is absent) [VESELOVSKY, VSHIVTSEV, 1988, P. 112].

Biotesting of environmental oil pollution

Biotesting identifies the environment toxicity by means of test objects standard reactions under controlled (laboratory) conditions. Its essence is detecting deviations of physiological mechanisms that ensure the homeostasis in living systems (at the molecular, cell and organism levels) in response to hostile conditions that cause stress [MELEKHOVA O.P., YEGOROVA E.I., 2007]. This helps detect when even the smallest deviations appear in the state of the studied environment within short time intervals.

An extensive set of methods is used for biotesting the oil pollution in water, soil, and sediments. This set includes practically all current approaches to detecting the organism response to their habitat pollution.

Morphological approach is based on identifying deviations (compared to baseline) in test objects morphology, their growth rate, and mortality.

It is common to determine the toxicity of oil-contaminated environments using higher plants as test objects: cultivated oat (*Avena sativa L.*), common wheat (*Triticum vulgare L.*), radish (*Raphanus sativa L.*), garden cress (*Lepidum sativum Linn*), dill (*Anethum graveolens L.*), mustard (*Brassica juncea (L.) Czern.*), white trefoil (*Trifolium repens L.*), and bulb onion (*Allium cepa L.*). These plants demonstrate a clear correlation between the germination capacity of seeds, the weight and length of germinants, and the plant height and branching on oil concentration in the soil. Significant deviations of these parameters from baseline values in many plants occur when the oil mass content in the soil exceeds 1%.

Soil and aquatic invertebrates are also widely used for oil pollution biotesting. The criterion of the polluted environment acute toxic effect is the mortality of 50% of specimens or more within a specified time and their ability to reproduce [MELEKHOVA O.P., YEGOROVA E.I., 2007].

Biochemical approach is used to assess the environment effect on living systems in terms of enzymatic activity and efficiency of biochemical reactions. Soil enzymes activity is determined by the amount of substrate processed under optimum conditions or the resulting reaction product [MELEKHOVA O.P., YEGOROVA E.I., 2007]. Most often, the catalase activity is determined, as it is strongly inhibited during soil pollution by oil and petroleum products.

Biophysical approach is based on instrumental detection of disruptions in physical and biochemical processes of test organisms due to the impact of external factors. Biotesting of oil-contaminated water, soils, and sediments allows for a wide use of the fluorometric and luminescent methods that ensure high sensitivity and allow make express tests of the environmental condition.

Fluorometry measures the intensity of chlorophyll fluorescence in photosynthetic organisms at the time of the photosynthetic apparatus switching from active into inactive state. This allows assessing the intensity of photosynthesis process. Microalgae from the *Chlorella* genus are most often used to test environmental quality with this method.

Electrical orientation test is based on detecting the electrical orientation effect (EOE) of bacterial cells exposed to oil hydrocarbons-containing medium as compared to baseline. EOE is measured by the changes in optical density of cell suspension under the influence of alternating electric field.

To assess the level of seawater pollution by oil and petroleum products, it is common to use the filtration ability of *Mytilus galloprovincialis mussels* [OSTROUMOV, 2005]. Biotesting using this method detects a reduction in the rate of polluted water filtration by mussels (by reduced optical density of water in the vessels where mussels are placed).

Physiological approach is based on assessment of physical values of the test organisms condition and their behavioral reactions to environmental pollution.

Biotesting of oil pollution in various environments is completed using the test with single-cell green algae *Chlorella vulgaris Beiyer* characterized by high sensitivity to oil hydrocarbons. The following parameters are detected in the cells that have grown in nutrient media for 24 hours: survival rate (evaluated by established increase of non-viable cells which do not form microcolonies on agar media), number (in Goryaev chamber), number of dead cells (discolored, lysing, with disfigured form), chlorophyll a content (determined by photocolorimetry), oxygen productivity (by Winkler method) [MELEKHOVA O.P., YEGOROVA E.I., 2007].

The practice of oil pollution biotesting also makes use of animal behavioral reactions to the oil and petroleum products presence in the environment. One of the tests is based on the reaction of maxillopods from the *Cladocera* order. In the presence of oil and petroleum products, they surface to the water-air interface [LOZOVOY, 2003]. It is established that the sensitivity of *Cladocera* surfacing behavioral reaction grows with increasing boiling temperature of oil and petroleum products, while in case of aromatic and aliphatic hydrocarbons it also grows with the increase of their molecular weight. For some petroleum products, this test is by 3–7 orders of magnitude more sensitive than the biotests based on the maxillopods survival rate.

Among other test objects are *Daphnia magna, Daphnia pulex, Ceriodaphnia pulchella G.O. Sars, Bosmina longirostris O.F. Müller, Simocephalus vetulus O.F. Müller, Chydorus sphaericus O.F. Müller*.

For biotesting of water, sediments, and soils in oil field areas, the Spirostomum infusoria (*Spirostomum ambiguum Ehrbg.*) is used as the test organism. Spontaneous motor activity serves as the integral indicator of the infusoria condition. The registered parameter is the number of times that the infusoria crosses the hairline in the binocular microscope ocular in 1 minute [MELEKHOVA O.P., YEGOROVA E.I., 2007].

Genetic approach is used to identify the environment's stress condition or toxicity through the degree that genetic changes (mutations) are manifested in the test object. Genetic tests are characterized by their rapid results (they can be obtained within a maximum of 2–3 weeks) and by high sensitivity to the presence of mutagenic substances in the environment. Levels of PAHs content in soils are biotested using the Ames test with the *Salmonella typhimurium* strains and the SOS chromotest with *Escherichia coli* [MARCHENKO ET AL., 2013]. The Ames test assesses mutagenic activity by the obtained number of reversed mutations to a prototrophic state in terms of histidine consumption [MELEKHOVA O.P., YEGOROVA E.I., 2007].

Comprehensive test systems

In biotesting of oil-contaminated environments, it is reasonable to apply a set of tests that constitute a test system. In this case, the advantages of some tests will make up for the limitations and drawbacks of others. This helps provide a more accurate integral assessment of the natural environment condition. One should use the following principles when choosing the methods of biotesting: the test system must consist of test objects that are different in their taxonomic position and constitute different links of the trophic chain. Test objects must be highly sensitive to oil hydrocarbons and must be quite easy to cultivate. Recorded parameters must be meaningful and, at the same time, must not require complex and expensive equipment. When developing a test system, one should take into account the specific characteristics of water, soils, sediments, and industrial waste under test: total salts quantity and their composition, alkali-acid and oxidation-reduction conditions, humic acid content, etc.

Comprehensive test systems use the following tests to indicate hydrocarbon pollution in various environments: bioluminescent test on photobacteria; photocolorimetric test of microorganisms reduction activity (which characterize the reaction of intracellular enzymes to pollution); electric orientation and osmo-optical tests on various bacteria (which identify changes in barrier properties of cell membranes); and growth tests on bacteria, microalgae, and higher plants [KHOLODENKO ET AL., 2001].

CHAPTER 14
STUDY OF HYDROCARBON GEOCHEMICAL
FIELDS IN THE SOIL COVER

14.1. TYPOLOGY AND ORIGIN
OF HYDROCARBON GEOCHEMICAL FIELDS IN SOILS

Concept of hydrocarbon geochemical field in soils

Matter flows related to exploration, production, and transportation of hydrocarbons come from a number of sources. If no information about the sources is available, it is not always possible to correctly assess the impact of the flows on the environment and to select the right practical environmental measures. Collecting this information is one of strategic areas of oil and gas geoecology. To that effect, the concept of "hydrocarbon geochemical field in soils" has been introduced [PIKOVSKY ET AL., 2008, 2012].

Geochemical fields of hydrocarbons in soils are the areas characterized by certain proportions between major classes of hydrocarbons. This concept is fundamentally different from most other notions of geochemical or geophysical fields (a geochemical field of a certain individual element, temperature field, etc.). The complexity of hydrocarbon fields in soils is specified by their multicomponent nature. Each of such fields is characterized by its own geochemical processes and its own combination of natural gases, bitumoids, polycyclic aromatic hydrocarbons (PAHs), and normal alkanes that are present in the soil and vary in qualitative composition and quantitative proportions. The nature of a hydrocarbon geochemical field provides the most comprehensive information on the origin and transformation stage of carbonaceous substances in the soil.

Diagnosis of hydrocarbon geochemical fields in soils helps identify hydrocarbon sources and processes that result in the formation of specific hydrocarbon geochemical anomalies.

The following genetic groups of geochemical hydrocarbon fields in soils are defined: (1) biogeochemical, (2) emanational, (3) atmo-sedimentational, and (4) impact. Each of these groups is characterized by its specific hydrocarbon sources of both natural and anthropogenic origin, formation processes, and unique composition — the content and ratio of different hydrocarbon components in soils (Table 14.1).

Table 14.1.

Types of hydrocarbon geochemical fields in soils
[Pikovskiy ET AL., 2012, P. 243]

Genetic group of hydrocarbon geochemical field	Processes and hydrocarbon sources		Examples of hydrocarbon geochemical fields in soils
Biogeochemical	natural	natural processes of organic residue decomposition	biogeochemical associated with humus lipids
			biogeochemical associated with peat lipids
	anthropgenic	organic components on landfills	biogeochemical anthropogenic
Emanational	natural	natural gas emanation from the Earth's interior	deep emanational
	anthropgenic	emanations of subsoil anthropogenic hydrocarbons	subsurface-emanational
Atmo-sedimentational	natural	aerosols settling after volcano eruption, natural fires	volcanogeic fire-induced
	anthropgenic	settling of solid aerosols from industrial plant, flare unit, heating equipment, vehicle emissions	atmo-sedimentational, dispersed
			atmo-sedimentational concentrated
Impact	natural	natural petroleum occurrences on the Earth's surface	natural oil and gas show on surface
			hydrothermal springs, mud volcanism
	anthropogenic	impacts of oil and gas production, transportation and oil refineries on the environment	oil spills
			petroleum products spills

Biogeochemical hydrocarbon fields in soils

A biogeochemical hydrocarbon field is the field of hydrocarbons created by biogeochemical processes of their synthesis. A biogeochemical field of hydrocarbons in soils is specified by low concentrations of gaseous hydrocarbons (mainly, methane), bitumoids, and polycyclic aromatic hydrocarbons. Components of soil humus or peat serve as nutritious substrates for hydrocarbon-producing microorganisms. The composition of biogeochemical field polyarenes in soils is determined by low-condensed homologues of naphthalene, fluorene, and phenanthrene that comprise 70–10% of the total identified PAHs.

In the studied soddy-podzolic soils (Glossic Retisols), odd homologues C_{25}–C_{33} dominate over even homologues in the composition of normal alkanes. Natural gases concentration in the composition of soil air is typically less than 200 mg/m^3. Methane is the primary component of gases.

Substantial differences in parameters of the natural hydrocarbon state of the soils in one climatic zone are observed between zonal and azonal (swamp peat) soils. Peat soils exhibit higher levels for all of the soil hydrocarbon complex components.

Emanational hydrocarbon fields in soils

Various types of emanational hydrocarbon fields result from underground emanations of hydrocarbons into soils. Hydrocarbons migrate from petroleum deposits (diffusion, effusion, cross-flow between layers), from human-made underground oil and petroleum products reservoirs, as well as from underground anthropogenic hydrocarbon lenses and flows.

Local and linear types of deep emanational hydrocarbon geochemical fields are identified by the ratio of gaseous hydrocarbon phase and other parameters. They characterize the aureole of deposit dispersion and hydrocarbon migration through tectonic faults, correspondingly. Furthermore, the subsurface-emanational hydrocarbon field is identified. It acts as an indicator of subsoil pollution — presence of anthropogenic hydrocarbon lenses or intense petroleum flows emerging as a result of deep fluid outburst from petroleum deposit to the surface.

Atmo-sedimentational hydrocarbon geochemical fields in soils

An atmo-sedimentational hydrocarbon geochemical field is formed due to fallout of solid aerosols on the soil surface. These aerosols contain hydrocarbons and other carbonaceous substances that are often quite toxic. Atmo-sedimentational hydrocarbon geochemical fields can be local and regional. Local fields are formed due to fallout of industrial emissions and combustion products of gas and condensate from the atmosphere. Regional fields are associated with regional and global anthropogenic transport of pollutants in the atmosphere resulting in their fallout on the soil surface at a long distance from the source. This source cannot always be identified. Regional atmo-sedimentational fields along with biogeochemical fields create the background against which the anomalies caused by natural or anthropogenic processes are revealed.

Impact hydrocarbon fields in soils

Surface hydrocarbon injections into the soil caused by oil and petroleum products spills or surface shows of natural hydrocarbons combine into local injection oil hydrocarbon fields of different dimensions. An injection oil field is detected by high values of natural gas and PAH concentrations, significant quantity of lube-resinous type bituminous substances, and presence of sulfurous compounds. Injection hydrocarbon fields can be divided based on the degree of oil pollutant degradation that is determined by the intensity of free natural gas showings, as well as by the qualitative composition of soil bitumoids.

Background hydrocarbon geochemical field

In natural landscapes that are not changed by anthropogenic processes, regional and local hydrocarbon fields of biochemical origin are identified. These hydrocarbon fields comprise the basis of soil hydrocarbon background. However, they are not the only source of background hydrocarbons. The level of hydrocarbon concentration in background soils depends on fallout of aerosols containing carbonaceous substances from the flows of global or regional transfer of pollutants in the atmosphere. Another source that contributes to the background level of hydrocarbon concentration is dispersed hydrocarbon subsoil degassing that is particularly noticeable in petroleum-bearing areas where hydrocarbon background exceeds the background outside petroleum-bearing territories by an order of magnitude.

14.2. HYDROCARBON GEOCHEMICAL FIELDS IN SOILS AT PETROLEUM PRODUCTION SITES

Indicative geochemical properties of hydrocarbon geochemical fields of various origins in soils of oil fields

The study of hydrocarbon geochemical fields in soils at operational facilities revealed their reliable role as indicators helping determine the origin of specific hydrocarbon groups in soils.

There are a number of distinctive genetic groups of hydrocarbon geochemical fields in an oil field. They are characterized by genetic homogeneity and similar hydrocarbon state of the soils (Tables 14.2 and 14.3, Fig. 14.1).

170

Table 14.2.

Examples of Parameters of anthropogenic and natural-anthropogenic hydrocarbon geochemical fields in oil field soils (southern taiga)
[KRASNOPEYEVA, 2009; PIKOVSKY ET AL., 2012, P. 251]

Type of hydrocarbon geochemical field and hydrocarbon source	Natural gas content	Bitumoid content and type	PAH content and association type	N-alkanes content and CPI*
Biogeochemical, regional (zonal soils)	Low	Low, light	Low, phenanthrene-naphthalene	Low, 4.4–9.6
Biogeochemical, local (peat soils)	Elevated	High, light with pigments	low to elevated, tetraphen-naphthalene	High, 10.8–12.4
Deep emanational (aureoles of deposit dispersion, emanations via tectonic faults)	High (400–1,000)	Low, light	Low, phenanthrene-naphthalene	Low, 2.4–7.2
Subsurface-emanational (hydrocarbon emanations from anthropogenic subsoil lenses)	High, heavy hydrocarbons	Low, light hydrocarbon, lube oil	Elevated, benzo[ghi] perylene	Low, 3.4–5.0
Injection, oil, weakly degraded (recent oil spill)	High, sulfurous gases	High, lube-resinous	High, benzo[ghi] perylene	High, 1.8–3.4
Injection, oil, degraded (old oil spill)	Low	High, resinous	High, benzo[ghi] perylene	N/a

* CPI (carbon preference index) is an indicator that describes the ratio between concentrations of n-alkanes with an odd and even number of carbon atoms in the molecule in the particular homologous series interval, from C15 and higher. It reveals whether the soil hydrocarbon composition is closer to oil or to plant lipids. The CPI for oil is between 1.0 to 1.2. For C_{24}–C_{32} n-alkanes extracted from higher plants it is 4–7, extracted from peat soils it is above 10.

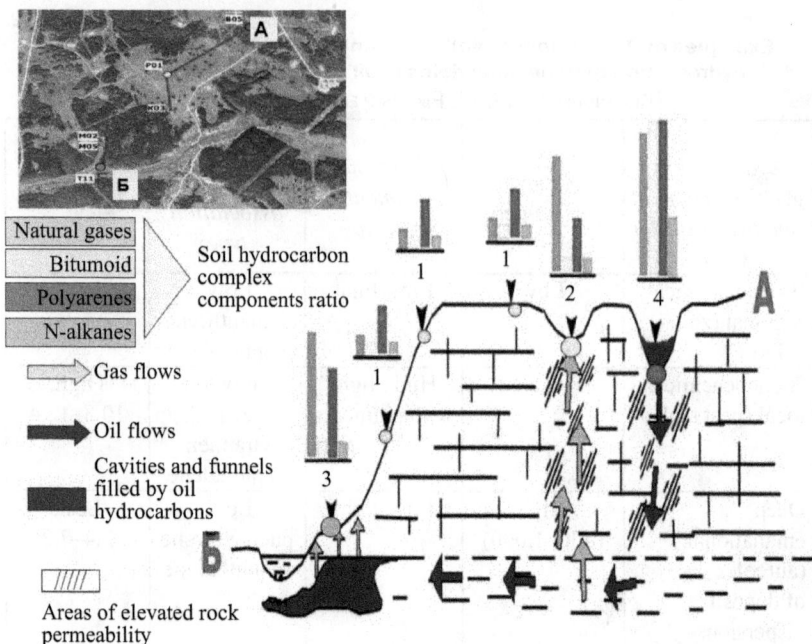

Fig. 14.1. Scheme of hydrocarbon geochemical fields (hydrocarbon states of the soils) in an oil field (karst landmass, Perm Region) [Pikovsky et al., 2012, p. 252]. Hydrocarbon states of the soils: (1) biogeochemical (background), (2) deep emanational associated with migration of hydrocarbons via tectonic faults, (3) subsurface emanational associated with subsoil lenses of oil and petroleum products, and (4) injection field associated with an oil spill on the surface.

In certain cases, it is reasonable to distinguish between the degraded subtype of injection hydrocarbon field in soils related to old oil spills the weakly degraded subtype related to fresh (or renewed) oil spills. At the sites of old oil spills, substantial soil degassing has taken place. Therefore, their hydrocarbon state is characterized by the low or moderate gas phase concentration, and, at the same time, by the significant content of oil-type bituminous substances and PAHs. At the sites of fresh oil spills, particularly in the soils of river valleys and footslopes, the maximum concentration of hydrocarbons in the gas phase and maximum soil pollution by bituminous substances and PAHs were recorded. Natural emanations of natural gases rising from the oil deposit to the Earth's surface through rock fracturing zones are identified based on criteria for a deep emanational hydrocarbon geochemical field in an uncontaminated

172

area. This type of hydrocarbon fields is characterized by substantially increased content of gaseous hydrocarbons (mainly, methane) in the soil atmosphere (up to 1,000 mg/m3) while maintaining background concentrations of bitumoids and PAHs and light compounds predominating in the composition of polyarenes that contain up to 4 rings.

Another type of emanational hydrocarbon field in soils is distinguished based on high contents of free hydrocarbons in soil and subsoil air, whereby the concentration of heavy methane homologues (C4-C6) in their composition sharply increases. At the same time, the concentration of bitumoids and PAHs in upper horizons of the soil profile remains at the background level, while in lower horizons (at a depth of 60–100 cm) both components occur in elevated quantities. The fraction of heavy hydrocarbons in the gaseous phase composition amounts to 60%. Among them, high contents of iso-pentane, n-pentane, and n-hexane have been recorded. These properties are indicative of a field of anthropogenic hydrocarbon emanations formed due to the impact of hydrocarbon vapor and gas flow rising from heavily polluted subsoil horizons or secondary oil accumulations in karst cavities and fractured rocks. This type of hydrocarbon geochemical field in the soil cover has been termed subsurface-emanational. A subsurface-emanational hydrocarbon field in the soil cover is an indicator of subsoil pollution.

Oil fields block structure, occasional intensification of the tectonic activity, disruption of a natural reservoir regime by prospective and production drilling, occurrence of fluid reservoirs with abnormally high and abnormally low formation pressures (AHFP and ALFP, respectively) create prerequisites for huge volumes of gas, liquid hydrocarbons, and water being redistributed between layers.

Therefore, it is of great importance to detect the deep-subsurface emanational geochemical field resulting from hydrocarbons migration via natural and human-made fractures in rock masses or through the annular space in wellbores. Fluids flowing through these channels result in formation of shallow anthropogenic lenses of oil, flammable gas, gryphon occurrences, pollution of the atmosphere, soil, water-bearing layers.

These phenomena intensify along with the tectonic activity which results in natural fluids making use of freshly formed natural fracturing zones. Gas-liquid flows are also provoked by powerful explosions in wells to create human-made reservoirs.

The history of petroleum field development is replete with examples of catastrophic accidents on deep wells caused by creating permeable channels and formation fluid and gas leaking to the surface. In any operational oil and natural gas field, particularly a large one, it is essential to deal with the issues of early (pre-accident) diagnosis of the zones where formation fluids actively leak to the surface, studying their behavior in the near-surface zone, predicting potential consequences of cross-flows, and taking timely measures to prevent and mitigate adverse phenomena. Appearance of gryphons and potential emergency overflowing of aggressive hydrogen sulfide gas can result in an environmental disaster. That is why it is necessary to implement a system to monitor fluid streams from gas-saturated zones to the surface, promptly detect the occurrence of hydrocarbon, hydrogen sulfide, and other anomalies against the natural geochemical background and of natural channels that can be used for fluid cross-flow from the depth.

Dynamic and static features of hydrocarbon geochemical fields in the areas of oil and gas deposits

Two groups of geochemical indicators can be distinguished among informative characteristics of hydrocarbon geochemical fields:

(1) *dynamic,* i.e., free gases and other mobile substances in soils, sediments, atmosphere, and water;

(2) *static*, i.e., indications of the migration of hydrocarbons in soils and sediments in the form of stable hydrocarbon compounds adsorbed on mineral surfaces, or traces of secondary rock changes.

The former include volatile hydrocarbons and non-hydrocarbon gases and mercury. Dynamic indicators can also include the thermal flow value. If the gas flow from the depth is temporarily cut off or modified, the dynamic anomaly can disappear or relocate.

Static indicators reflect long-term fluid streams, whose substance passes into an immobile state. Gaseous components can be subjected to biogeochemical or catalytic transformation into light bituminous substances. This results in new mineral formations and changes of organic substance, for example, formation of polycyclic aromatic hydrocarbons. Static indicators help reveal long-living migration channels, i.e. the most dangerous sites that must be monitored.

The informative value of dynamic and static indicators depends on whether the conditions of certain component formation correspond to the surface conditions. Among dynamic indicators, the following are of informative value: heavy saturated and unsaturated hydrocarbons in soil and subsoil air, mercury vapor, hydrogen, radioactive emanation products, and sulfurous gases. Among static indicators, polycyclic aromatic hydrocarbons (PAHs) in the composition of epigenetic bituminous matter of soils and sediments are highly informative.

174

The above-mentioned indicators act as direct indicators of the fluid flowing to the surface or of anthropogenic pollution. The possibility of human-induced technological contamination can be excluded via taking samples from the depths of 5–10 meters and analyzing the potential sources of substance inflow and the patterns of its spatial distribution.

CHAPTER 15
ENVIRONMENTAL MONITORING
OF PETROLEUM ACTIVITIES

15.1. PURPOSE AND TYPES OF ENVIRONMENTAL MONITORING

Environmental monitoring is the main method of nature conservation activity. This is a multi-purpose system designed to track the environment condition and the way it changes due to the impact of natural and anthropogenic factors.

The aim of environmental monitoring is to assess and preemptively predict the potential negative effects of several groups of factors on humans and the environment:

(1) natural factors (earthquakes, tsunami, volcano eruptions, avalanches, mudslides, climatic anomalies, etc.);

(2) natural-anthropogenic factors (condition and impact of geotechnical complexes — engineering structures and natural objects that function together as a whole: wells, mines, dams, etc.);

(3) anthropogenic factors (the status and environmental impact of engineering structures, production facilities, vehicles, etc.; process and emergency discharge of polluting substances, mechanical disturbances of the soil cover and vegetation, accidents at hazardous production facilities, etc.).

Environmental monitoring efficiency depends on its feedback, i.e., on using its data to optimize production facilities and operating conditions, control emissions volume, prevent emergencies, and optimize other areas of human activity to be safer for the environment.

Environmental monitoring types are also classified in terms of natural media that are being monitored (atmosphere, hydrosphere, soil, ecosystem, lithosphere), natural environment condition (background, emergency), set of applied methods (aerospace, ground, subsurface), the substances being monitored and other properties.

All types of environmental monitoring can be divided into three large groups depending on their functional purpose:
- evaluation monitoring;
- operational monitoring;
- emergency response monitoring.

The task of evaluation monitoring is to periodically (up to once a year) assess and predict the environmental condition at various territorial levels in order to take measures for minimizing the negative trends. This monitoring is performed at the level of countries, regions, districts, and local areas involved in the economic activity. The objects of evaluation monitoring are atmospheric air, soils, surface water (including hydrobiological indicators), ozone layer of the atmosphere, ionosphere, and near-earth space. Evaluation monitoring criteria are the limits specified for maximum allowable emissions and concentration of polluting substances. It is common practice to monitor a large group of substances that circulate in this territory due to production peculiarities and all environment components — atmosphere, surface and underground water, bottom sediments, snow cover, soils, vegetation, and animal life. Main methods of substances' diagnosis must strictly comply with the appropriate regulatory limits.

Operational monitoring is used to continuously (or at short time intervals) monitor the environment for hazardous natural or anthropogenic phenomena, including spontaneous or industry-provoked emissions and discharges of hazardous polluting substances, mechanical disturbance or subsidence of the surface, and anthropogenic changes in the atmosphere and hydrosphere. The monitoring is aimed at preemptively predicting hazardous impacts on the environment from natural and anthropogenic groups of factors in order to take early measures to prevent the environmental hazard.

The main criterion of operational monitoring is the change in a local background level of monitored substances concentration that had been established in this territory before the observations began. Operational monitoring is executed as part of the local production monitoring covering the footprint of one or several production enterprises that are interconnected by their location. The aim of the monitoring is to identify the environment signals from an incipient contamination or disruption of environmental components or other environmentally hazardous natural phenomena that are about to appear. A small number of substances that are of high priority for this branch of industry are monitored in the media that are the most sensitive to increased concentration of monitored substance.

Monitoring the results of emergency response measures — oil spills and other major accidents, as well as the results of polluted land remediation to remove residual pollution by oil and petroleum products, includes elements of both operational and assessment types of monitoring. On the one hand, it involves tracking the rapidly changing situation as part of the contamination response and, on the other hand, monitoring is finished only when the specified regulatory level of environmental pollution is ensured.

15.2. EVALUATION ENVIRONMENTAL MONITORING OF PETROLEUM ACTIVITIES

All enterprises of the petroleum industry implement production environmental monitoring aimed at ensuring compliance with national and regional environmental law. Petroleum exploration and production is monitored by petroleum companies at their licensed sites.

It is not an easy task to organize environmental monitoring at a petroleum-producing complex (i.e. the territory combined into a single natural and industrial cluster) where natural, natural-anthropogenic, and purely anthropogenic factors are tightly interwoven. Oil and gas field construction, well drilling, petroleum production, transportation, treatment, and processing result in intensification of vertical (from the interior) and lateral (in landscapes) hydrocarbon flows, processes of natural environment degradation, and deterioration of human living and working conditions. The monitoring is aimed at maintaining the balance between production and man's natural habitat.

Industrial monitoring functions, firstly, as a system for observations, assessment, and forecasting of natural environment condition; secondly, as local (operational) monitoring that provides a system for "assessing the environmental situation and human activity consequences within limited areas of natural systems that are exposed to direct technical impact, for example, in the course of operating wells, pipeline, etc." [KAZHDOYAN YU.S., KASIMOV N.S., 2008, P. 315].

Environmental regulation of landscape component pollution is the main criterion for environmental assessment, which is the benchmark of assessment environmental monitoring.

Environmental pollution regulation includes two aspects. The first one is the maximum allowable discharge of oil, gas, and petroleum products from process facilities into the environment. The second once is the maximum allowable concentration of pollutants in natural environments. If this concentration is exceeded, the environment is considered to be polluted.

The norms for maximum allowable pollutant emissions into the atmosphere and water bodies, as well as waste generation norms have been developed in the greatest detail [BUKHGALTER ET AL., 2003]. Regulation of emergency oil and petroleum product discharge into soils is profoundly different from regulating industrial enterprises emissions. For the latter, the maximum allowable discharge into air and water bodies is specified. Many substances, including hydrocarbons, are subject to maximum allowable concentration norms specified for water and air.

178

It is impossible to develop general unified norms of land pollution by oil and petroleum products because soils and vegetation respond to the same pollution levels differently depending on the natural conditions and the composition of pollutant. Besides, all soils have their own hydrocarbon background created by biogeochemical processes in the soil itself and by regional atmospheric fallout.

The criteria of soil pollution by substances of hydrocarbon origin can be the state of soil, plants, and soil animals, as well as the results of biological objects testing. Soil buffering depends on its resistance to pollution and the pollutant's composition. Therefore, hazardous levels of oil and petroleum product concentration in soils will vary significantly depending on the region.

The threshold level of oil and petroleum product content in soils and sediments above which the quality of natural environment will deteriorate is regarded as the maximum safe level of oil and petroleum product concentration in soils [Pikovskiy, 1993]. This level depends on the combination of numerous factors, such as productivity change or signs of soil toxicity, deterioration of vegetation state, groundwater pollution, etc.

It is not sufficient to specify only the maximum safe level of petroleum product concentration to regulate soil pollution. Natural ecosystems have a huge potential for natural purification. They included intensive physico-chemical and microbiological processes that degrade hydrocarbons. Therefore, if the source of contamination is promptly identified, petroleum product concentration in the soil will decrease naturally, ultimately reaching the safe level. It is not reasonable to perform special soil remediation activities because of the ever present danger to cause even greater harm to the soil ecosystem. One should specify a level of petroleum product concentration above which the soil is no longer able to cope with the pollution. This concentration level could be referred to as the limit of soil natural purification potential.

Soils that contain petroleum product concentrations above the natural purification limit are subject to sanitization and remediation because, if these measures are not taken they will never leave the degradation stage and will continue producing a stable negative effect on the environment.

The quantitative approaches to regulating the petroleum product content in soils that are implemented in a variety of countries are close to this concept and specify different levels for soils contamination. The choice of environment preservation measures depends on these levels. The norms take into account the nature of regional environmental pollution, industrialization rate, environmental policy, physiographical conditions that facilitate or prevent natural purification of the environment.

Table 15.1

**Tentative allowable levels for concentrations
of petroleum products of different compositions
in soils of major natural zones of Russia**
[PIKOVSKY ET AL., 2003, P. 1139]

Substances	Soil groups	TAC (including background concentrations), mg/kg	Impact on organisms
Light petroleum products: gasolines, kerosene, diesel fuel	a) Tundra gley loamy and clay; tundra swamp (Oxiaquic Cryosols, Histic Cryosols);	2,000	Short-term strong narcotic effect, inhibition of microbiological and photosynthetic activity
	b) Middle and southern taiga podzols and sandy loamy podzolic and soddy-podzolic soils (Podzols, Retisols, Luvisols (Arenic));	4,000	
	c) Southern taiga soddy-podzolic loamy (Glossic Retisols (Siltic));	4,000	
	d) Forest-steppe and steppe gray forest, chernozems (Phaeozems, Cher-nozems);	8,000	
	e) Brown desert-steppe (Calcisols);	8,000	
Heavy petroleum products: crude oil, residual fuel oil, lube oils	a) Tundra gley loamy and clay; tundra swamp (Oxiaquic Cryosols, Histic Cryosols);	700	Deterioration of the soils hydrophysical properties, deceleration of pho-tosynthetic activity, carcinogenesis
	b) Middle and southern taiga podzols and sandy loamy podzolic and soddy-podzolic soils (Podzols, Retisols, Luvisols (Arenic));	2,000	
	c) Southern taiga soddy-podzolic loamy (Glossic Retisols (Siltic));	2,000	
	d) Forest-steppe and steppe gray forest, chernozems (Phaeozems, Cher-nozems);	4,000	
	e) Brown desert-steppe (Calcisols).	2,000	

Based on the generalized global practice and experimental data, safe concentration of oil and petroleum products in different type of soils vary between 1,000 and 5,000 mg/kg. The concentrations between 5,000 and 20,000 mg/kg are referred to as moderate and enable the soils self-purification within three years. Serious damage when remediation is required starts from the level of oil content in the soil at 20,000 mg/kg and higher [McGill, 1977].

The concept of "tentative allowable concentration" (TAC) can be used to develop the official regulatory system. TAC of soil pollution by petroleum products is understood as the lowest allowable pollution level which allows the soil to restore its productivity in one year under given natural conditions, and at which negative consequences for the soil biocenosis can be naturally mitigated. Based on experimental research in several regions of Russia, Azerbaijan, and Ukraine, TACs for oil and petroleum products were proposed for major types of zonal soils (Table 15.1). This understanding of TAC as a general sanitary indicator can be applied for the upper humus-accumulative soil horizon (approximately to the depth of 20–30 cm). Data on the entire soil profile can be used to develop the water migration hazard indicator, i.e., the ability of chemical substance to move from the soil into groundwater and surface water sources.

There are still not enough experimental data to specify petroleum product TACs for different natural zones. The proposed TACs are somewhat lower than the data obtained in certain areas. It is quite safe to assert that petroleum product concentrations in the soil below the levels given will not cause any noticeable harm to the environment, and the corresponding hydrocarbon quantities will be almost completely retained in the upper soil horizons. In a year, these concentrations will be substantially lower and, since no other toxic compounds are present in the composition of petroleum products, the latter will not affect soil productivity and toxicity.

15.3. OPERATIONAL ENVIRONMENTAL MONITORING OF PETROLEUM PRODUCTION COMPLEXES AND MAIN OIL AND GAS PIPELINES

Operational environmental monitoring of licensed sites is aimed at solving the following tasks:

- give warning if negative changes occur in any of the main components and complexes in the oil field territory by quickly identifying the source of oil and petroleum products discharge into the environment and provide urgent information to top management of the enterprise and environmental agencies in case of a sharp increase in level of pollutant content in natural environments;

- ensure efficient recovery of polluted and damaged land by monitoring emergency response measures and their results (large spills of oil, petroleum products, saline water, explosion and fire consequences), as well as monitor how residual soil pollution by oil, petroleum products, and saline water is mitigated;
- perform technical monitoring.

The practical procedures of operational environmental monitoring have not been yet sufficiently developed both in the methodical and regulatory aspect. This is one of the topical tasks of oil and gas geoecology.

Operational environmental monitoring of petroleum production

The technique of early warning about the danger of environmental pollution by oil, gas, and petroleum products is based on receiving continuous remote quantitative information on levels of the most informative pollution indicators in various zones of the licensed sites. This information can appear in transit media that transport pollutants: atmosphere, surface, and groundwater.

The atmosphere can be monitored using standalone methane and total hydrocarbon sensors that continuously send data to the central point.

'Aquatic environment monitoring can be provided by installing standalone chloride and luminescent oil film sensors. Surface water sensors are installed on permanent water flows, for example, in small river estuaries. Subsurface water monitoring requires drilling observation wells where the same sensors are installed.

Formation fluid leaks from wells are detected by increased chloride concentration and by the luminescent film in water. Oil from tanks or damaged pipelines will quickly reveal itself by oil films appearing on water and by increased total hydrocarbon concentration in the ground-level atmosphere. Natural gas flows will be detected by the sharp increase of methane content. A quick response from the emergency team of environmental specialists will ensure that the point of hydrocarbon leaks is identified and possible accidents are prevented. If the necessary sensors are not available or impossible to use, a daily drive through monitoring points is organized where water and air tests are performed using field analyzers.

Operational industrial monitoring of main oil and gas pipelines

A main pipeline is under permanent danger of being damaged. Normally, it is laid in the active soil layer exposed to seasonal phenomena. An artificial drain runoff is created along the pipeline. Besides, the pipeline is exposed to external geodynamic impacts, particularly strong in earthquake prone areas.

182

The aim of operational monitoring on a functional main gas, oil, or petroleum product pipeline is to detect fluid leaks early enough to contain them. Often, small leaks resulting from pipe wear and damage can rapidly evolve into accidents, explosions, and fires.

Continuous monitoring of linear pipeline systems includes four main sets of methods:

- remote monitoring using continuously updated information from satellites, aircraft, and helicopters. It includes interpreting data from multi-zonal and spectral zonal, radar, thermal, and other types of surveying, identifying anomalies in the imagery so that the emergency research team could be dispatched for their analysis;
- setting up stationary observation points in the areas of highly intensive landscape-geochemical and geological processes that can cause pipeline deformation, damage or rupture. It is crucial to establish these points in active geodynamic zones — at the joints of modern crustal block boundaries [Rantsman, Glasko, 2004]. These areas are responsible for the largest number of pipeline accidents. Block boundaries often follow large river valley which crossings represent high hazard sections. One must also pay attention to pipeline sections laid in lowland, poorly ventilated areas where explosive gases that leak from the pipe can accumulate;
- organize monitoring observations of fluid composition at compression stations to promptly detect aggressive components that contribute to pipe corrosion and toxic components in gas (mercury, hydrogen sulfide);
- organize regular surface monitoring of a gas state of the soil using field gas analyzers in-between stationary observation points. Where surface observations are impeded, special attention should be paid to remote observation of changes in vegetation condition.

15.4. EMERGENCY RESPONSE MONITORING

Terrestrial emergency response monitoring

Emergency oil and petroleum products spill monitoring is performed simultaneously with the activities to deal with emergency consequences. Its main task is to provide the personnel with timely information required for taking urgent measures on spills cleanup and ensure that they are implemented efficiently and with the minimal damage to the environment. The monitoring is performed continuously throughout the duration of emergency measures.

Primarily, the monitoring measures employing field diagnosis methods must provide the following information:
- spill source location, approximate volume of spilled oil or petroleum product, and their qualitative description (density, viscosity, flash and freezing temperature, toxicity);
- area and volume of contamination, whether its boundaries can expand over time;
- oil flow direction and travel speed, possible depth of oil penetration in soil areas with different resilience;
- ground and surface water pollution hazard in the area of contamination;
- explosion, fire, and terrain gas contamination hazard;
- forecast of weather conditions and dangerous natural phenomena;
- hazard level for environmental components and living conditions of people.

The task of the monitoring is to track changes in the situation throughout the entire period of such activities, predict its development, and update the incoming information.

Monitoring of the polluted land remediation process

The following information is required to draw up a plan for land purification from residual oil:
- quantity and composition of oil and water-soluble salts in the soil horizons;
- state of vegetation on the polluted area;
- soil agrochemical and agrophysical properties (moisture content, pH, dissolved solids, content of carbon, nitrogen, potassium, phosphorus, carbonates);
- monitoring of the soil remediation process (changes in a hydrocarbon state and toxicity of the soil, its content of water-soluble salts, humus, biocenosis recovery, growth of sown annual and perennial plants, condition of ground and surface water in remediation areas).

Monitoring of the emergency response to marine environment pollution

Most often, the reason why an emergency monitoring system of the marine environment is deployed is oil spill accidents. Urgent answers to questions on spill spread, geochemical behavior of pollutants, and spill impact on ecosystems are given by the set of urgent monitoring measures. In most oil-producing countries, they are an indispensable component of national and corporate plans for prevention and elimination of oil spills and their consequences.

The monitoring of an emergency oil and petroleum product spill consists of the two phases.

1. Phase of direct observation of the spill nature and spread. Spill source location, the nature of spilled substance, spill volume, and accurate time are identified.

2. Phase of assessing the environmental impact of the spill.

Aerial observations are a crucial element of the polluted area monitoring program. When flying over polluted areas, one should pay attention to movement paths of pinnipeds and other mammals within the spill and in its vicinity, as well as to migration of birds and turtles that could come into contact with oil.

Post-spill environment condition monitoring is primarily aimed at identifying oil hydrocarbon concentrations in bottom sediments and biota.

When results must be obtained quickly, complex analytical procedures are not normally used.

Fluorometry remains a fairly efficient method of oil spill monitoring. Oil contamination on the sea surface can be identified by means of fluorescence lidars capable of not only detecting oil contamination around a sea object but can also estimate the oil film thickness, which makes it possible to give a quantitative assessment of pollution.

Examples of emergency spill monitoring due to shipwrecks and tanker accidents in the coastal zone are the most numerous.

The problem of identifying petroleum product sources can sometimes be an extremely sensitive one if several emergencies occur simultaneously. To solve this problem, oil samples from tankers and other potential pollution sources are analyzed.

CHAPTER 16
FORECASTING HAZARDS AND RISKS
OF ENVIROMENTAL CHANGE DUE TO THE IMPACT
OF PETROLEUM ANTHROPIZATION

16.1. TASKS AND METHODOLOGY
OF ENVIRONMENTAL FORECASTING

Due to the global nature of petroleum anthropization, forecasting its impact on the environment becomes the cornerstone problem of oil and gas geoecology. The task of geoecological forecasting is to predict potential areas of dangerous environmental changes due to the impact of petroleum production within the forecast land or water territory.

The level of hazard related to natural environment modification by petroleum production, transportation, storage, and processing can be indicated by the following groups of factors:
* condition and resilience of the environment and its components (atmosphere, soils, waters, biocenoses) to anthropization and its response to anthropogenic impact under specific natural conditions;
* Earth's crust geodynamic activity creating risks of operation faults in process facilities — wells, pipelines, buildings, and constructions;
* the nature of potential disruption of environmental components by the industrial activity, quantity, and composition of the substance extracted from the interior, volume, and composition of produced and consumed petroleum products.

Forecasting methodology is based on territorial zoning depending on the hazard level of petroleum production impact on the environment. An object position in the zoning system determines the potential risks and methods of protection against the negative impact of petroleum anthropization. Environmental resilience and responses are covered by the landscape-geochemical forecast, while the geodynamic activity of the Earth's crust is covered by the geodynamic forecast.

Landscape-geochemical forecasting

Landscape-geochemical forecast maps are based on original landscape-geochemical maps developed in Russia by M.A. Glazovskaya and A.I. Perelman.

Forecast landscape-geochemical zoning according to M.A. Gla-zovskaya consists in distinguishing the two territorial units: cascade landscape-geochemical systems (CLGS) and technobiogeoms.

Cascade landscape-geochemical systems are high-order basins of liquid and solid runoff that include lower-level basins — landscape-geochemical arenas (LGA) interconnected by the substance runoff. Various CLGS and LGAs are differentiated by the pattern of substance migration and the sites of their transit or final accumulation (see Chapter 11).

Technobiogeoms are the spatial aggregates of natural landscapes that possess a similar level of geochemical resilience and similar pattern of environmental responses to anthropization (see Chapter 8). Technobiogeoms are identified within the boundaries of lower-order LGAs and vary in terms of their natural purification rate and the nature and rate of substance transformation. A feature of forecast landscape-geochemical zoning is that it is focused on modern processes and factors that define the way a landscape-geochemical system functions and the level of its resilience to anthropization. Assessment of the effects of anthropization on the natural environment and the choice of protective measures takes into account natural factors of landscape self-purification.

Geodynamic forecasting

Modern geodynamic activity of the Earth's interior is one of the most important factors that affect petroleum production operations. Oil and gas fields inside Earth's crust are closely linked to recent tectonic structures in terms of their location. Geodynamic activity of the interior manifested by seismic and microseismic phenomena affects the integrity of process facilities on fields and pipelines. Its influence shows in sudden surges of reservoir pressure, spontaneous petroleum blowout at oil wells, process facilities deformation often resulting in emergency situations and severe environmental consequences. M.A. Glazovskaya highlighted seismicity as one of the risk factors of natural environment changes at petroleum production enterprises and main oil pipelines. This risk is particularly high in mountainous and piedmont landscapes. However, it exists even on extensive plains where strong earthquakes never occur. It is associated with barely noticeable microseismic fluctuations of the Earth crust in the areas of elevated modern geodynamic activity. These fluctuations are the most intense on the boundaries of modern Earth's crust tectonic blocks, particularly, on the intersection of the boundaries of three or more blocks — in disjunctive nodes. The territory block structure is identified through the link between modern relief and its morphostructure and recent movements of Earth's crust (GERASIMOV, 1993; MESCHERYAKOV, 1981).

The method of identifying modern block structure of Earth's crust and its most active areas was developed by Russian geomorphologists E.Ya. Rantsman and M.P. Glasko for mountainous and plain territories. They prepared a number of forecast maps in various scales for the territory of Russia, European countries, North and South America, and certain Asian countries (RANTSMAN, GLASKO, 2004). The method is based on morphostructural zoning (MSZ) of territories based on formalized informative relief properties. The main properties are the relief heights and linear elements ratios, as well as the nature of river valley patterns. Based on spatial homogeneity of these properties within a certain territory, blocks of various hierarchy levels are identified — the areas that are the most stable in geodynamic terms. Where this homogeneity is disrupted, the transition to another block is observed. Block boundaries (morphostructural lineaments) are manifested by intensified linear zones varying in width. Intersection or junction areas of block boundaries of different directions are identified as morphostructural nodes that correlate with disjunctive tectonic nodes. Morphostructural nodes are the most active areas of Earth's crust and are very dangerous for the process facilities located there. In the nodes, not only surface but also deep processes intensify: Earth's crust tectonic fragmentation contributes to more intense vertical migration of water and gas flows. These flows are fixed in geochemical anomalies, which is typical of many nodes.

MSZ diagrams are hierarchically structured cartographic models of the modern Earth's crust block structure. Model elements constitute a unified system and cannot be considered in isolation from each other.

In seismically active zones, the nodes correlate with the strongest earthquakes, and in petroleum-bearing basins — with large hydrocarbon accumulations.

In relatively "calm" regions, the nodes are associated with spontaneous accidents on process facilities. These include ruptures of main gas and oil pipelines, railway line deformations, suddenly collapsed large constructions, failures of gas and oil distributors, and various compressors. Underground gas and petroleum product reservoirs constructed in the nodes are the least reliable storage of liquid and gaseous substance.

Therefore, landscape-geochemical forecast maps predicting natural environment changes under the petroleum production impact are to be supplemented by forecast maps of active geodynamic phenomena. These can result in spontaneous emergency situations aggravating the anthropogenic impact on the environment. Geoecological forecasting is not limited to the two types of zoning described above. Regional and local forecasts could require, for example, climatic, biocenotic, and other possible types of forecast zoning that are still to be developed.

16.2. FORECAST LANDSCAPE-GEOCHEMICAL ZONING OF THE TERRITORY OF RUSSIA AND NEIGHBORING COUNTRIES IN TERMS OF NATURAL ENVIRONMENT MODIFICATION BY OIL PRODUCTION

Zoning tasks and forecast units

Combined forecast landscape-geochemical zoning of the territory of Russia and neighboring countries in terms of the types of potential natural environment modification by oil production was completed at the department of landscape geochemistry and soil geography of the Faculty of Geography at the Lomonosov Moscow State University [GLAZOVSKAYA, 1988; 2007; GLAZOVSKAYA ET AL., 1983; PIKOVSKIY, 1993]. In the following years, the zoning was updated for the territory of Russia [NATIONAL ATLAS OF THE SOILS OF RUSSIAN FEDERATION, 2011; ECOLOGICAL ATLAS OF RUSSIA, 2017]. Combined zoning was supplemented by compiling stability maps of individual environment components (soils, water, and vegetation) for the same territories. These maps were the basis of forecast zoning for the entire territory of the former USSR [PROBLEMS OF GEOGRAPHY..., 1983, NATIONAL ATLAS OF RUSSIA, 2007; GENNADIEV, PIKOVSKY, 2007].

The task of combined zoning was to typify landscape-geochemical arenas and technobiogeoms in terms of stability and risk of potential landscape changes due to mechanical disruptions, and spills of oil and field wastewater.

Zoning elements on the map include forecast landscape-geochemical areas (CLGS of the 1st order) and forecast landscape-geochemical districts (technobiogeom). Forecast landscape-geochemical areas were differentiated by the final place of anthropization products accumulation and were characterized by the direction and volume of runoff from the basin territory. Forecast landscape-geochemical districts were identified within the boundaries of forecast areas, mainly within small and medium river basins. Districts similar in stability and landscape character were united in types, and the types similar in physico-geographical conditions were united in groups of landscape-geochemical districts. The forecast for possible changes of the environment was provided for types and groups of landscape-geochemical districts.

The nature of anthropogenic load from oil production was determined for each landscape-geochemical district and district subtypes. District subtype layer was plotted on the map in terms of produced fluid composition — the combination of oil and reservoir water.

Territorial delineation by positions of final pollutant accumulation
(forecasting landscape-geochemical areas)

Overall, 26 landscape-geochemical areas, which are essentially the arenas of runoff into the seas of the Arctic, Pacific, and Atlantic oceans, into the Caspian and Aral seas, and into smaller endorheic basins in Central Asia were identified in the former Soviet Union. Each landscape-geochemical area has a common runoff to positions of the final accumulation of anthropization products (Fig. 16.1). Location of pollution sources in the cascade landscape-geochemical system becomes especially important for describing the conditions that enhance or obstruct the environment ability to withstand the anthropogenic impact of petroleum production in this area. Petroleum production could be present in the territory of the forecast area in whole or in part. In some cases, petroleum production districts can be located in the middle and lower parts, in other cases in upper and middle parts, and, finally, in the lowest parts (deltas, estuaries) of the cascade system. This determines the distribution of contaminants across the area's territory, their dilution, or thickening.

Territorial delineation by types of environmental changes
(forecast landscape-geochemical districts)

A forecast landscape-geochemical district (LGD) is identified if it belongs to one type of landscape and one drainage basin of the second or third order.

Types of forecast landscape-geochemical districts reflect similar groups of geological factors specifying geoecological risk for modification of the landscape. Each type is characterized by a specific combination of soil and climatic conditions that determine similar environmental responses to anthropogenic impact. District types are differentiated by intensification or weakening of certain natural processes, position of geochemical barriers, rate of landscape natural purification and recovery. There are 27 types of forecast landscape-geochemical districts. Each district is characterized by its unique scenario of landscape components change caused by mechanical disturbance of soils; destruction of vegetation; and spills of oil, petroleum products, and effluent reservoir water on the soil surface.

The district types with similar physico-geographical conditions are united into five groups. Each group is characterized by certain common forms of possible anthropogenic changes in the environment caused by the landscape-geochemical processes that dominate in their territory: transformation, dispersion, and secondary thickening of contaminants — oil and mineralized field water (Table 16.1).

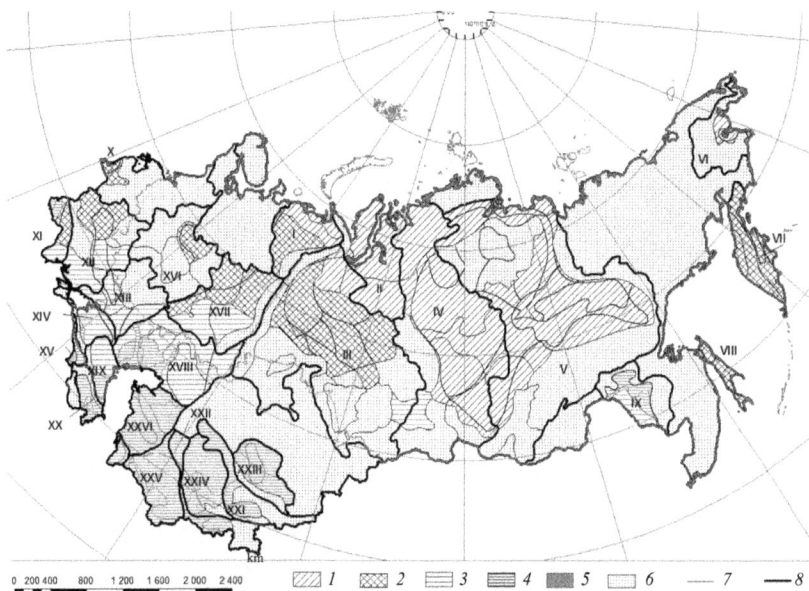

Fig. 16.1. Forecast zoning of Russia and neighboring countries by types of potential environmental changes related to petroleum production [GLAZOVSKAYA ET AL., 1983] Technobiogeom groups (forecast landscape-geochemical districts): (1) permafrost tundra and taiga, (2) taiga-forest, (3) steppe, (4) semidesert and desert, and (5) subtropical forest; (6) territories without of oil production; (7) boundaries of technobiogeom types. Forecast landscape-geochemical areas (pollutant drainage basins): (8) boundaries of landscape-geochemical areas. *Areas of discharge to the Arctic Ocean:* I — Ural-Timan; II — Yamal-Gydan; III — West Siberian; IV — Central Siberian; V — East Siberian. Areas of discharge to the Pacific Ocean: VI — Chukotka-Anadyr; VII — Kamchatka; VIII — Sakhalin; IX — Amur River region. *Areas of discharge into the seas of Atlantic Ocean:* X — Baltic; IX — Cis-Carpathian; XII — Dnieper River region; XIII — Donetsk-Azov; XIV — West Cis-Caucasian; XV — West Trans-Caucasian. *Areas of runoff to the Caspian Sea and intermediate reservoirs of the basin:* XVI — Upper Volga; XVII — Volga-Kama; XVIII — Caspian; XIX — East Cis-Caucasian; XX — South Caspian. *Areas of runoff to the Aral Sea:* XXI — Fergana-Syr Darya; XXII — Amu Darya. *Areas of runoff to continental deltas and closed drainage depressions:* XXIII — Karatau; XXIV — Turkestan-Kyzyl Kum; XXV — Kopet Dagh-Kara Kum; XXVI — Ustyurt-Mangyshlak.

Table 16.1

Forecast scenarios of environmental changes under the impact
of oil production and transportation operations in different groups
of landscape-geochemical districts [GLAZOVSKAYA, 1988]

Groups of forecast landscape-geochemical districts	*Description of natural zones*	*General characteristics of environmental changes related to oil production*
Permafrost-affected tundra and taiga	Permafrost area. Widespread occurrence of swamps and wetlands, as well as dissected high plains and plateaus. Areas of arctic, typical and southern tundra, forest-tundra and woodlands, middle-taiga and south-taiga landscapes	All forms of cryogenesis: thermokarst, thermal erosion, frost heave, solifluction; mechan-ical migration of pollutants. Long-term accumulation of oil on geochemi-cal barriers, in water bodies and bottom sediments, development of hy-drogen-sulfide conditions and barriers, high level of plant degradation, intensification of gleyzation processes, soil salinization, concentration of salts and non-volatile hydrocarbons in waters, low rate of natural purification
Taiga forest areas	Taiga forest and broadleaved forest zones beyond permafrost area. Widespread occurrence of swamps on lowland plains	Accumulation of oil products on reducing geochemical barri-ers; development of hydrogen-sulfide barriers. Relatively slow minerali-zation of petroleum products in soils and water bodies, their settling in sedimentation basins; favorable conditions for migration with surface and ground waters, dilution and dispersion; temporary salinization with sub-sequent soil alkalization and soil swamping
Steppe areas	Subarid and arid regions where evaporation exceeds precipitation, with a high radiation balance and limited or absent surface runoff	Relatively fast decomposition of oil products, evaporation of volatile oil fractions; accumulation of heavy stable oil components in the bottom sediments of lakes, reservoirs, and river floodplains; increased risk of soil salinization in depressions and enclosed water bodies; mechanical surface disruptions create a hazard of sheet and ravine erosion and deflation

Groups of forecast landscape-geochemical districts	Description of natural zones	General characteristics of environmental changes related to oil production
Semi-desert and desert areas	Subarid and arid regions where evaporation significantly exceeds precipitation, with a high radiation balance and limited or absent surface runoff	Intense soil deflation caused by mechanical disturbance; development of eolian landforms. Active decomposition of petroleum product in wetted soils and on surface of water bodies; settling of heavy oil fractions on the bottom of water bodies, intensive soil salinization by field water resulting in secondary stable saline lands. In case of heavy oil contamina-tion, binding and cementing of soil particles, formation of hydrogen-sulfide barriers is possible in water bodies and hydromorphic soils
Mountain areas	Diversity of landscape-geochemical settings that vary with the absolute height	Mechanical and intensive water migration of contaminants. Rapid mineralization of organic pollutants, intensive re-moval from soils and dispersion of soluble substances in surface water. Local accumulation of petroleum products and sulfurous compounds at reducing barriers in seashore deltas

16.3. FORECASTING EMERGENCY SITUATIONS IN THE AREAS OF HIGH GEODYNAMIC ACTIVITY

Forecasting maps of morphostructural zoning (MSZ) were prepared for several regions of Russia (RANTSMAN, GLASKO, 2004). The study focused on the link between geodynamic phenomena and emergency incidents at petroleum transportation facilities and the elements of the block structure of these territories: blocks, block boundaries, and morphostructural nodes. An example is given using the MSZ diagram for the central and southern parts of the Russian plain (Fig. 16.2).

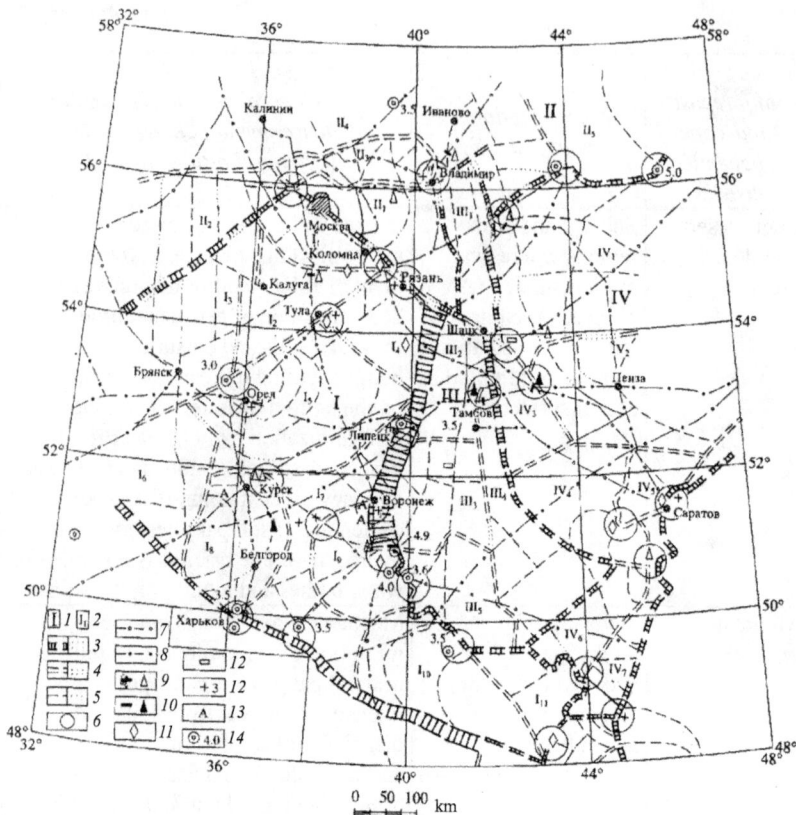

Fig. 16.2. Location of engineering facility emergencies on the map of the modern Earth's crust block structure in the center and south of the Russian plain [Rantsman, Glasko, 2004, p. 159].

1, 2 — Block hierarchy: *1* — first rank — macroblocks; *2* — second rank — mesoblocks; *3–5* — morphostructural lineament zones: *3* — first rank; *4* — second rank; *5* — third rank (dash lines if clearly expressed in surface forms; dot lines if not clearly expressed); *6* — conventional border of morphostructural nodes (*25* km radius) with known emergencies; *7, 8* — main pipelines: *7* — gas pipelines; *8* — oil pipelines; *9–13* — emergencies: *9* — on gas pipelines; *10* — on oil pipelines; *11* — on railroads; *12* — sudden collapses of permanent constructions, *13* — involving industrial utility lines, oil distributors, production and storage facilities of highly toxic substances (HTS), *14* — nuclear power plants; *15* — epicenters of earthquake with a magnitude above *3–5.*

194

On the MSZ maps for the center and south of the Russian Plain and Cis-Caucasian Region, all epicenters of platform earthquakes with a magnitude of 3–5 are located in the morphostructural nodes zone 25 km in radius. In 1989–1994, in the center and south of the Russian Plain, Cis-Caucasian region, and Western Siberia, there were 88 emergency incidents on oil and gas pipelines, engineering constructions, railways, compressor stations, and other facilities, including 92% in morphostructural nodes with the same radius. Among them, out of 29 emergencies on oil and gas pipelines, 9.6% proved to be in morphostructural nodes (Table 16.2).

Table 16.2.

Distribution of emergency situations in 1989–1994 within the block structure of the Earth's crust according to MSZ maps

| Emergency type | Region | Total emer-gen-cies | Including | | | Total emer-gencies | Emer-gencies in block nodes |
			within blocks	on block boun-daries	in nodes, within the radius of 25 km		
Accidents on oil and gas fields	Russian Plain	11	1	2	8	29	26
	Cis-Caucasia	10			10		
	Western Siberia	8			8		
Sudden collapses of buildings and constructions	Russian Plain	4	1		3	5	4
	Cis-Caucasia	1			1		
Railroad accidents	Russian Plain	11	1		10	17	16
	Cis-Caucasia	6			6		
Emergencies at oil and gas distributors, compressor stations, poison-ous substance reservoirs, etc	Russian Plain	24	1		23	37	35
	Cis-Caucasia	8			8		
	Western Siberia	5	1		4		

CONCLUSION

Oil and gas environmental ecology is a new independent field of science, which is still being developed. The major patterns of the behavior of hydrocarbons in the biosphere are already known. However, many problems remain unsolved. Let us summarize our review of the fundamental issues of oil and gas geoecology.

Oil and petroleum products are a multitude of substances that differ in composition, properties, and environmental functions. Oil in general is not a toxic product. In some circumstances and doses, it has a favorable effect on living organisms. Individual components in its composition can be toxic. These include aromatic hydrocarbons, sulfurous compounds, and certain metals if their concentrations in the oil solution are high enough. Many natural associates of petroleum are also toxic (mercury, hydrogen sulfides, radionuclides, highly mineralized reservoir water) and can have an even stronger impact on environmental components than oil itself. Studying the composition of oil and its associates in the areas of its production and occurrence is one of the main tasks of petroleum geoecology. An important challenge in this field is the study of natural hydrocarbon flows in the environment.

Negative effect of petroleum anthropization on soils, vegetation, surface and subsurface terrestrial waters, and aquatic ecosystems of the World Ocean depends on the particular natural conditions; production methods; volumes of substance inflow, outflow, and dispersion; and the composition of anthropization products. Petroleum extraction from the lithosphere is no longer limited to discrete deposits. Today, the industry employs unconventional petroleum resources that are dispersed in shale and sand masses stretching across tens and hundreds of thousand square kilometers (shale gas, shale oil, oil sands). It is not yet quite clear what the costs are for the environment, but it is already evident that in this case they are much higher than with conventional petroleum production.

All measures related to mitigation of pollution consequences and remediation of damaged land must be grounded on the main principle: not to cause any more harm to the ecosystem than the harm caused by pollution. Polluted land remediation should be understood as an extension of the natural purification process aimed at its acceleration using natural purification and adaptation mechanisms. Any concept for polluted ecosystem recovery must be grounded on this principle. Its essence consists in the maximum mobilization of internal ecosystem resources in order to restore its original functions. Selection of land remediation technology must be based on soil resilience to oil pollution. Environmental regulation principles of soil pollution by oil and petroleum products and polluted land remediation technologies can be justified based on the assessment of natural purification mechanisms.

There are no universal methods to clean up oil pollution. Getting to know the general laws of soil and aquatic ecosystem recovery and the ways to apply such laws to specific natural conditions is a vital task that is highly relevant for preservation of the biosphere in general.

Diagnosis of pollution and other disturbances in the environment is one of the key issues of oil and gas geoecology. If it is not adequately resolved, it is impossible to reliably assess the environment condition, its compliance with environmental norms, and even to draw up these norms. However, a consistent analytical concept of the object of diagnosis is still absent from regulatory documents. Petroleum products are most often understood as just the total hydrocarbons—substances extracted from soils and water by non-polar or low-polar organic solvents and then stripped of polar components. The most dangerous high molecular weight substances in this case are either disregarded completely or excluded from "petroleum products." This situation undermines the importance of environmental monitoring because the information on environmental condition and, first of all, soils, becomes incomplete and, hence, distorted.

To increase the informative value of geoecological study of hydrocarbons in the environment, the notions of "hydrocarbon state of the soils" and "hydrocarbon geochemical field" have been introduced. These notions describe the whole complex of soil hydrocarbons: natural gases, bituminous substances, and individual hydrocarbon compounds with due account for their distribution patterns in the soil profile and landscape-geochemical system. Diagnosis of a hydrocarbon state of the soils gives a more complete picture of the soil and bottom sediments pollution level, as well as of the origin of carbonaceous substances, thus helping us to differentiate between the natural and anthropogenic aspects of carbonaceous components. Study of hydrocarbon geochemical fields as a spatial aggregate of homogeneous hydrocarbon states of the soils is regarded as a promising method of identifying processes that result in the development of certain hydrocarbon anomalies in soils.

Environmental pollution by petroleum production is not limited to oil and petroleum products. Dozens of associated substances can have a toxic effect on ecosystem components. It is unrealistic to perform the diagnosis of each substance and understand the nature of its impact on living organisms. Methods are required to perform the complete ecosystem diagnosis at a given concentration of major pollutants. These

197

methods include bioindication and biotesting that are now widely applied for environmental monitoring. Bioindication assesses the environmental condition via analyzing both organism communities and indicator species populations, as well as individual indicator organisms directly in their habitat. Biotesting consists in detecting deviations of physiological mechanisms that ensure the homeostasis in test objects under reference conditions (at molecular, cell, and organism levels).

Monitoring and forecasting of the state of environment under the impact of petroleum industry are required to take timely measures for protecting the environment from degradation. Environmental monitoring includes three aspects. The first aspect is assessment monitoring based on periodic environmental quality assessment and the pattern of its change related to production activities. Under different climatic conditions, the environment exhibits varying resilience to the impact of anthropogenic substances. Therefore, different hygienic norms should be applied in different landscapes and natural zones. The second aspect of environmental monitoring is operational monitoring. Its task is to prevent emergency situations and take early measures to avoid damage to the environment and production facilities by continuously monitoring changes in the environmental components located within the footprint of the functioning production complex. The third aspect is continuous monitoring of the response to emergency incidents: oil spills during transportation and storage, oil and gas blowouts due to uncontrolled flowing, fires, explosions, and other negative events.

An efficient monitoring system can be introduced only based on a geoecological forecast of possible negative events in the environment. The tasks of forecasting main types of environmental changes under the impact of petroleum anthropization are solved by the methods of geoecological zoning of territories and water bodies. Forecast landscape-geochemical zoning is the division of any territory according to the tolerance of its landscapes towards anthropization and the risk of changes in the environmental components under mechanical or chemical impacts. Morphostructural zoning builds a model of the Earth's crust block structure, identifies local areas with the highest potential activity of the Earth's crust activity, where the highest risk of deformation and destruction of process facilities is expected. Other types of forecast zoning may also exist for elaboration of geoecological forecasting.

Oil and gas environment ecology has yet to establish its methodology and the definitive set of tools. It will go on developing, absorbing new facts and new areas of environmental research, and tackle the challenges that the human civilization sets for society.

REFERENCES

Abraham H. (1945) Asphalts and allied substances. Their occurrence, modes, of production, uses in the Arts and methods of testing. Vol. one. Raw materials and manufactured products. — New York, D. Van Nostrand Company, Inc.

Abrosimov A.A. (1999) Ecological aspects of production and application of petroleum products. Moscow, BARS publishing house, 1999. (in Rus)

Abrosimov A.A. (2002) Ecology of hydrocarbon systems processing: Textbook. Moscow, Chemistry publishing house, 2002. (in Rus)

Alekseeva T.A., Teplitskaya T.A. (1981) Spectrofluorometeric methods of analysis of aromatic hydrocarbons in natural and anthropogenic environment. Leningrad, Gidrometeoizdat publishing house (in Rus)

Alyakrinskaya I.O. (1966) On behavior and filtration ability of Mediterranean mussels in oil-contaminated water. Zool. zhurn., 1966, 45: 998–1003

Artemieva T.I. (1989) Groups of soil animals and the problems of remediation of anthropogenic territories. Moscow, Nauka publishing house

Artemieva T.I., Zherebtsov A.K., Borisovich T.M. (1988) The influence of contamination of soil by petroleum and oil field waste water on the group of soil animals. In: Glazovskaya M.A., ed. Remediation of oil-contaminated soil ecosystems. Moscow, Nauka publishing house

Aske N., Kallevik H., Sjöblom J. (2001) Determination of saturate, aromatic, resin, and asphaltenic (SARA) components in crude oils by means of infrared and near-infrared spectroscopy. Energy Fuels. 15:1304–1312

Belyaev S.S. (1988) Methane-forming bacteria: biology, systematics, application in biotechnology. Successes of microbiology 22:169–206 (in Rus)

Beskrovny N.S. (1993) Rational ways of reclamation of traditional and non-traditional raw hydrocarbon resources (on the basis of the international experience). Saint Petersburg, Braun-Blanquet J. Pflanzensociologie. Wien — New York, 1964. — p. 865

Bukhgalter E.B., Golubeva I.A., Lykov O.P. et al. (2003) Ecology of oil and gas sector. Moscow, Neft & Gaz Publishing house (in Rus)

Butuzova G.Yu. (2003) Hydrothermal sedimentary mineralization in the global ocean. Moscow, EDU publishing house (in Rus)

Buzmakov S.A., Kostarev S.M. (2003) Anthropogenic changes of environmental components in the oil producing areas of Perm region. Perm, Perm university publishing house (in Rus)

Danilov A.M. (2003) Introduction into chemmotology. Moscow, Tekhnika publishing house, TUMA GRUPP LLC (in Rus)

De Lima C.G. (1986) The Shpol'skii effect as an analytical tool // CRC Crit. Rev. Analyt Chem. 1986. V. 16. № 3. P. 177–221

Dedyukhina E.G., Zheliphonova V.P., Yeroshin V.K. (1980) Hydrocarbons of microorganisms// Successes of microbiology 15: 84–98

Dobryansky A.F. (1961) Chemistry of oil. Leningrad, Gostoptekhizdat publishing house, 1961

Dorokhova M.F. (2008) The state of environment in the petroleum production area in accordance with the state of soil microbiota. In: Kazhdoyan Yu.S., Kasimov N.S., eds. Oil and environment of Kaliningrad region. Vol. 1, Land. M.-Kaliningrad, Yantar Skaz publishing house, p. 190–205

Dorokhova M.F., Solntseva N.P. (2012) Experimental researches of the migration processes of oil in soil of Kaliningrad region. In: Geochemistry of landscapes and soil geography. In commemoration of 100 years of M.A. Glazovskaya. Moscow, APR publishing house, 2012, pp. 259–276

Ecological atlas of Russia (2017) Moscow, Pheoria publishing house

Ehhalt D.H. (1974) The atmosphere cycle of methane. Tellus 26:58–70.

Fan T. et al. (2002) Evaluating crude oils by SARA analysis society of petroleum engineers in: Proceeding of SPE. Improved Oil Recovery Symposium, April 13–17 (in Rus)

Florovskaya V.N., editing (1975). Luminescent bitumenology. Moscow, Moscow University publishing house

Frumin G.T. (2000) Ecological chemistry and ecological toxicology. Saint Petersburg, RGTMU publishing house, 2000. 198 p.

Galkin S.V. (2002) Hydrothermal biotic communities of the global ocean. Moscow, EDU publishing house

Gennadiev A.N., Pikovskiy Yu.I. (2007) The maps of soil tolerance toward pollution with oil products and polycyclic aromatic hydrocarbons: Methodological aspects. Eurasian Soil Science 40 (1): 70–81.

Gennadiev A.N., Pikovsky Yu.I., editing (1996) Geochemistry of polycyclic aromatic hydrocarbons in rocks and soils. Moscow, Moscow university publishing house

Gennadiev A.N., Pikovsky Yu.I., Tsibart A.S. and *Smirnova M.A.* (2015a) Hydrocarbons in Soil: origin, composition and Behavior (Rewiew). Eurasian Soil Science 48 (10): 1076–1089

200

Gennadiev A.N., Pikovsky Yu.I., Zhidkin A.P., Kovach R.G., Koshovskii T.S., Smirnova M.A., Khlynina N.I. and Tsibart A.S. (2015b) Factors and features of the hydrocarbon status of soils. 48 (11): 1193–1205

Glazovskaya M.A. (1988) Geochemistry of natural and anthropogenic landscapes of USSR. Moscow, Vysshaya shkola publishing house,

Glazovskaya M.A. (2007) Geochemistry of natural and anthropogenic landscapes. Moscow

Glazovskaya M.A. (1972) Technobiogeoms — original physico-geographical objects of landscape and geochemical forecasting. Vestnik Moskovskogo universiteta journal. Issue 5. Geography 6: 23–35

Glazovskaya M.A., ed. (1981) Anthropogenic substance flows in landscapes and condition of ecosystems. Moscow, Nauka publishing house

Glazovskaya M.A., ed. (1982). Extraction of mineral resources and geochemistry of natural ecosystems. Moscow, Nauka publishing house

Glazovskaya M.A., ed. (1983) Landscape-geochemical zoning and environmental protection. Problems of geography. Issue 120. Moscow, Mysl' publishing house

Glazovskaya M.A., ed. (1988). Remediation of oil-contaminated soil ecosystems. Moscow, Nauka publishing house

Glazovskaya M.A., Pikovskiy Yu.I. (1980) The rate of natural purification of soil from oil in various natural areas. Priroda journal, 1980, 5: 118–119

Glumov I.Ph., Kochetkov M.V., editing (1996) Anthropogenic contamination and the processes of natural purification of the Ciscaucasia area of the Black Sea. Moscow, Nedra publishing house

Goldberg V.M. (1998) Relation of groundwater and environment contamination. Moscow

Goldberg V.M., Zverev V.P., Arbuzov A.I. and others (2001) Anthropogenic contamination of groundwater with hydrocarbons and ecological implications. Moscow, Nauka publishing house

Gulieva S.A. (1981) Unique therapeutic Naftalan oil. Baku, Azerneshr publishing house (in rus)

Guzev V.S., Levin S.V., Seletsky G.I., Babieva E.N. and others (1989). The role soil microbiota in remediation of oil-contaminated soils. In: Microorganisms and soil protection. Moscow, Moscow university publishing house, p. 129–150

Harris, R.H. (1990) Ground Water and Soil Remediation: Conflicts and Opportunities. — In: Ground Water and Soil Contamination Remediation: Toward Compatible Science, Policy and Public Perception.

Report on a Colloquium Sponsored by the Water Science and Technology Board. Colloquium 5 of a Series. — Washington. D.C. National Academy Press, p. 19–37

Haymon R. M., Macdonald K.C. (1985) The geology of Deep-Sea Hot Springs. American Scientist, Vol. 73(5): 441–449

Isidorov V.A. (2001) Organic chemistry of atmosphere. Saint Petersburg, Khimizdat (in rus)

Isidorov V.A., Kondratiev K.Ya. (2001) Atmospheric methane. Ecological chemistry journal, 10 (13): 145–160

Ismailov N. (2006) Remediation of oil-contaminated soils and drilling sludge. Baku, Elm publishing house

Ismailov N.M. (1988) Microbiology and enzymatic activity of oil-contaminated soil. In: Remediation of oil-contaminated soil ecosystems. Moscow, Nauka publishing house p. 42–56.

Ismailov N.M. (2009) Practical ecotechnology. Baku, TPP "Tyakhsil" publishing house

Israfilov G. Yu., Listengarten V.A. (1978) Groundwater and Apsheron land reclamation. Baku, Elm publishing house

Kapustin V.M., Gureev A.A. (2007) Oil processing technology. Part two. Destructive processes. Moscow, KolosS publishing house

Kasumyan A.O. The impact of chemical pollutants on the feeding behavior of fish and its sensibility for feeding stimuli. Voprosy Ikhtiologii journal. 2001. Vol. 41, No. 1, pp. 82–95.

Kasymov A.G. (2004) Plankton ecology in Caspian sea. Baku, Adilogly publishing house

Kazhdoyan Yu.S., Kasimov N.S., editing (2008) Oil and environment of Kaliningrad region. Vol. 1, Land. M.-Kaliningrad, Yantar Skaz publishing house

Khaustov A.P., Redina M.M. (2006) Environmental safety on oil exploration areas. Moscow, Delo publishing house

Khlystov O.M., Gorshkov A.G., Egorov A.V. and others (2007) Oil in the lake of world's heritage. Doklady of Academy of Sciences, 414 (5): 656–659

Kholodenko V.P., Chugunov V.A., Zhigletsova S.K and others (2001) Development of biotechnological methods of elimination of environmental oil contamination. Ros. khim. journal 45(5–6):135–141

Kireeva N.A. (1995) Microbiological processes in oil-contaminated soils. Ufa. Bashkirsky State University publishing house

Kropotkin P.N., Valyaev B.M. (1981) Geodynamics of mud volcano activity (in relation to oil and gas content). From the book: Geological and geochemical bases for oil and gas exploration. Kiev, Naukova Dumka publishing house

Kustenko N.G., Podolyak G.P. (1982) The impact of oil on the stages of cellular cycle of two species of diatomic algae. Biologiya Morya journal 5:67–69

Kvenvolden, K.A. (1993) A primer on gas hydrate. In: Howell, D.G. et al (eds) / The Future of Energy Gases. US Geological Survey Professional Paper 1570. P. 279–291.

Kvesitadze G.I., Khatisashvili G.A., Sadunishvili T.A., Yevstigneeva Z.G. (2005) Metabolism of anthropogenic toxicants in higher plants. Moscow, Nauka publishing house

Link, W. (1952) Significance of oil and gas seeps in world oil exploration. AAPG Bul. 36 (8):1505–1540

Lukin A.E. (2011) The nature of shale gas in the context of problems of oil and gas petrology. Geology and mineral resources in the global ocean 3:70–85

Makovsky V.M. (1988) The impact of oil contaminants on the state of mire ecosystems in Surgut Priobye. In: Ecology of oil and gas sector. Moscow, p. 213–216.

Mamedov G. Sh., Ismailov N.M. (2006) Scientific bases and principles for zoning of Azerbaijan soil in accordance with the stability to contamination with organic substances. Baku, Elm publishing hous

McGill W.W. (1977) Soil restoration following oil spills — a review. J. Canad. Prtrol. Technol. 16 (2): 60–67

Melekhova O.P., Yegorova E.I., editing (2007) Biological control of environment: bioindication and biotesting. Study guide. Moscow, Academia publishing house

Melkov B.G., Sergeeva A.M. (1990) The role of hard carbonaceous substances in the formation of endogenic uranic mineralization. Moscow, Nedra publishing house

Mironov O.G. (1985) Interaction of sea organisms with oil hydrocarbons. Leningrad, Gidrometeoizdat

Mironov O.G. (1973) Oil contamination and sea life. Kiev, Naukova Dumka publishing house

Mironov O.G., editing. (1988) Biological aspects of sea environment contamination. Kiev, Naukova Dumka publishing house

Moskovchenko D.V. (2013) Ecogeochemistry of oil production areas of Western Siberia. Novosibirsk, Geo academic publishing house National atlas of the Russian Federation soil (2011) Moscow, Astrel, AST publishing house

National Atlas of the Russian Federation vol. 2. (2007)

Nemirovskaya I.A. (2004) Hydrocarbons in ocean (snow, ice, water, suspension, bottom sediments). Moscow, Nauchny mir publishing house

Nemirovskaya I.A. (2013) Oil in ocean (contamination and natural flows). Moscow, Nauchny mir publishing house

Neronov V.V. (2008) Vegetation response to oil production. In: Kazhdoyan Yu.S., Kasimov N.S., eds. Oil and environment of Kaliningrad region. Vol. 1, Land. M.-Kaliningrad, Yantar Skaz publishing house

Nesterova M.P. (1992) The bases of physical and chemical methods and means of fighting oil contamination of water environment. In: The modern problems of studying and protection of biosphere. Vol. 3. The problems of remediation and protection of biosphere systems. Saint-Petersburg, Gidrometeoizdat, p. 146–154

North W.I., Ir M.N., Clendenning K.E. (1965) Successive biological changes, observed in a marine cove, exposed to a large spillage of mineral oil. In: Pollutions marines par les microorganisms et les produits petroliés (Symposium de Monaco, 1964). Paris

Oborin A.A., Ilarionov S.A., Nazarov A.V., Khmurchik V.T., Markarova M. Yu. (2008) Oil-contaminated biogeocenose. Perm

Oborin A.A., Kalachnikova I.G., Maslivets T.A. (1988) Natural purification and remediation of oil-contaminated soil of Transurals and Western Siberia. In: Recovery of oil-contaminated soil ecosystems. Moscow, Nauka publishing house, p. 140–158.

Oboron A.A., Rubinshtein L.M., Khmurchik V.T., Churilova N.S. (2004) Organization conception of underground biosphere. Ekaterinburg, Ural Department of Russian Academy of Sciences publishing house

Orlov D.S. (1985) Soil chemistry. Moscow, Moscow University publishing house

Ostroumov S.A. (2004) Biological mechanism of natural purification in natural reservoirs and water streams: theory and practice. Success of modern biology journal, 124(5):429–442

Ozerova N.A. (1986) Mercury and endogenic mineralization. Moscow, Nauka publishing house

Patin S.A. (1997) Ecological problems of development of oil and gas resources of sea shelf. Moscow, the publishing house of Russian Federation Research Institute of Fishery and Oceanography

Perelman A.I., Kasimov N.S. (1999) Landscape geochemistry. Moscow, Astreya-2000 publishing house

Pikovsky Yu. I. (1993) Natural and anthropogenic hydrocarbon flows in environment. Moscow, Infra-M publishing house

Pikovsky Yu.I., Genadiev A.N., Oborin A.A., Puzanova T.A., Kras-nopeeva A.A., Zhidkin A.P. (2008) Hydrocarbon status of soils in an oil-producing region with karst relief // Eurasian Soil Science 41 (11): 1162–1170

Pikovsky Yu.I., Gennadiev A.N., Krasnopeeva A.A., Puzanova T.A. (2012) Natural and anthropogenic hydrocarbon geochemical fields in soils: conception, typology, indication value. In: Geochemistry of landscapes and soil geography. In commemoration of 100 years since the birth of M.A. Glazovskaya. Moscow, APR publishing house, p. 236–258

Pikovskiy Yu.I., Ismailov N.M., Dorokhova M.F. (2015) The bases for oil and gas geoecology. Moscow, INFA-M publishing house

Pikovskii, Yu.I., Korotkov L.A., Smirnova, M.A., Kovach R.G. (2017) Laboratory Analytical Methods for the Determination of the Hydrocarbon Status of Soils (a Review). Eurasian Soil Science 50 (10):1125–1137

Rantsman E.Ya., Glasko M.P. (2004) Morphostructural nodes — the places of extreme natural phenomena. Moscow, Media-Press publishing house

Reshetnikov A.I., Zinchenko A.V., Paramonova N.N. and others. Experimental researches of methane emissions in the natural gas production area on the northern part of Eastern Siberia. Ecological chemistry, 2006, 15 (3). pp. 147–166

Revell P., Revell Ch. (1995) Our habitat. Book 2. Water and air contamination. Moscow, Mir publishing house (in Rus)

Romanovskaya G.I., editing (2015) Luminescent analysis. Problems of analytical chemistry. Vol. 19. Moscow, Nauka publishing house

Rovinsky F.Ya., Teplitskaya T.A., Alekseeva T.A. (1988) Background monitoring of polycyclic aromatic hydrocarbons. Saint Petersburg, Gidrometeoizdat publishing house, 1988. 224 p.

Sanadze G.A. (1961) Plant volatile organic substances emissions. Tbilisi

Sedykh V.N., Ignatiev L.A, Semenyuk M.V. (2004) Plant reaction on drilling waste. Novosibirsk, Nauka publishing house

Shabad L.M. (1973) On circulation of carcinogens in environment. Moscow, Meditsina publishing house

Shiels W.E., Goering J.J., Hood D.W. (1973) Crude oil phytotoxicity studies. In: Environmental Studies of Port Valdez, p. 413–446

Shtina E.A., Nekrasova K.A. (1988) Algae from oil-contaminated soil. In: Glazovskaya M.A. ed., Remediation of oil-contaminated soil ecosystems. Moscow, Nauka publishing house, p. 57–81

Shvergunova L.V. (2000) Comparative and geographical analysis of the state of plants in oil production areas. Vestnik Moskovskogo universiteta journal, Issue 5, Geografiya (6): 28–33

Simoneit B.R.T. (1995) Organic geochemistry of water system at high temperatures and high pressure: hydrothermal oil. In: General branches of geochemistry. In the commemoration of 100 years since the birth of academician A.P. Vinogradov. Moscow, Nauka publishing house, p. 236–259

Sizykh V.I., Dzyuba A.A., Isaev V.P., Kovalenko C.N. (2004) Oil and gas problems of Baikal depression. Otechestvennaya geology, No. 5

Soboleva E.V., Guseva A.N. (2010) Chemistry of fossil fuels. — Moscow, Moscow University publishing house

Solntseva N.P. (1998) Oil extraction and geochemistry of natural landscapes. Moscow, Moscow University publishing house

Solntseva N. (2009) Environmental effects of oil production. APR, Moscow

Solovyev S.A. (2003) Natural gas hydrates as potential mineral product. Russian chemical journal, 47(3): 59–69

Stepanyan O.V., Voskoboinikov G.M. (2006) The impact of oil and petroleum products on morphofunctional characteristics of sea macroalgae. Sea biology journal 32(4): 241–248

Ugrekhelidze D.Sh. (1976) Metabolism of exogenic alkane and aromatic hydrocarbons in plants. Tbilisi, Metsniereba publishing house

Ugrekhelidze D.Sh. and *Durmishidze S.V.* (1984) Ingress and detoxication of organic xenobiotics in plants. Tbilisi, Metsniereba publishing house

Veselovsky V.A., Vshivtsev V.S. (1988) Biotesting of oil contamination of environment in accordance with the responses of plant photosynthetic apparatus. In: Glazovskaya M.A. ed. Remediation of oil-contaminated soil ecosystems. Moscow, Nauka publishing house, p. 99–112

Vinnem J.E. (1999) Offshore Risk Assessment. — Kluwer Academic Publishers. Dordrecht-Boston-London

Voskoboinikov G.M., Matishov G.G., Bykov O.D. and others (2004) On stability of sea macrophytes to oil contamination. Doklady Academy Nauk, 397(6): 842–844

Yakutseni S.P. (2005) Availability of hydrocarbon resources, enriched with heavy accompanying elements. The assessment of ecological risks. Saint Petersburg, Nedra publishing house

Yudakhin F.N., Gubaidullin M.G., Korobov V.B. (2002) Ecological problems of development of oil deposits in the northern part of Timan-Pechora province. Ekaterinburg

Zimonina N.M. (1998) Soil algae of oil-contaminated lands. Kirov publishing house

CONTENTS

www.ingramcontent.com/pod-product-compliance
Lightning Source LLC
Chambersburg PA
CBHW070926030426
42336CB00014BA/2561